Trailering Your Horse

A Visual Guide to Safe Training and Traveling

Cherry Hill

Photography by
Richard Klimesh

The mission of Storey Publishing
is to serve our customers by publishing practical information
that encourages personal independence in harmony with the environment.

Edited by Deborah Burns and Marie Salter
Copyedited by Doris Troy
Cover design by Eugenie Delaney
Cover photograph by Richard Klimesh
Text design and production by Susan Bernier
Production assistance by Jennifer Jepson Smith
Photographs and line drawings by Richard Klimesh, except where otherwise indicated; line drawings on pages 39, 42, 103, 118, and 130 by Alison Kolesar
Indexed by Susan Olason/Indexes & Knowledge Maps

Printed in Canada by Transcontinental Printing
10 9 8 7 6 5 4 3

Library of Congress Cataloging-in-Publication Data

Hill, Cherry, 1947–
Trailering your horse : a visual guide to safe training and traveling / Cherry Hill.
p. cm.
Includes index.
ISBN 1-58017-176-1
1. Horses—Transportation—Handbooks, manuals, etc. 2. Horse trailers—Handbooks, manuals, etc. I. Title.

SF285.385 .H55 2000
636.1'083—dc21

99-043744

Acknowledgments

Thanks to the following for their help with this manuscript:

American Horse Council, Washington, DC
B&W Custom Truck Beds, Inc., Humboldt, Kansas
Double Diamond Halter Co., Inc., Gallatin, Montana
Draw-Tite, Canton, Michigan
Equine Travelers of America Inc., Arkansas City, Kansas
Farnam Companies, Inc., Phoenix, Arizona
Featherlite Inc., Cresco, Iowa
Fold-A-Feeder, Big Piney, Wyoming
4-Star Trailers, Oklahoma City, Oklahoma
Tim Gueswel, La Porte, Colorado
Johnson Barns and Trailers, Phoenix, Arizona
Max-Air Trailer Sales, Fort Collins, Colorado
Scott Murdock Trailer Sales, Loveland, Colorado
Reese Products, Inc., Elkhart, Indiana
Lonny Smith, Cresco, Iowa
Sundowner Trailers, Inc., Coleman, Oklahoma
Turnbow Trailers, Oilton, Oklahoma

Also by Cherry Hill

101 Horsemanship and Equitation Patterns
Stablekeeping
Longeing and Long Lining the English and Western Horse
101 Longeing and Long Lining Exercises, English and Western
Beginning English Exercises
Intermediate English Exercises
Advanced English Exercises
Beginning Western Exercises
Intermediate Western Exercises
Advanced Western Exercises
Horse Health Care
Horse Handling and Grooming
Your Pony, Your Horse
Horse for Sale
101 Arena Exercises
Practical Guide to Lameness (with Dr. Ted S. Stashak)
Maximum Hoof Power (with Richard Klimesh)
Making Not Breaking
Becoming an Effective Rider
Horsekeeping on a Small Acreage
From the Center of the Ring
The Formative Years

Contents

✦ PREFACE ✦

We are a mobile society and we often want to take our horses with us when we go. Whether traveling to a weekend horse show or trail ride, moving across the country, or taking a horse for an emergency visit to the vet, you need to know safe trailering practices.

Safe trailering starts with a good rig — a suitable trailer and an adequate towing vehicle. An important key to low-stress trailering is spending time on training and familiarization lessons for your horse. For your actual trip, you'll need to pack the right gear and emergency equipment, learn how to drive while pulling a trailer, and know how to take care of your horse en route.

Whether you are a man or a woman, I really encourage you to become familiar with all aspects of trucks, trailers, training, and traveling: You will gain the knowledge, confidence, and experience to have a safe trip with your horse.

As my cover girl, I chose my high-mileage Quarter Horse mare, Zinger, aka Miss Debbie Hill (named after her sire, Smutty Hill, not me!). Zinger and I have done a lot of traveling together since 1975, when I purchased her as a yearling in Washington. I moved her from the coast to Washington State University, where I was doing a summer sabbatical, then back up to Alberta, Canada, where I had a full-time teaching position in the equine program at Olds College.

When my husband and I returned to the States in 1977, we had only a two-horse trailer, so we had to make two trips to haul our three horses and belongings to the Midwest. Sassy and Poco went on the first trip, which meant Zinger came second and shared the horse trailer with a bookcase and a filing cabinet in the adjoining stall. When normal winter weather turned into a wicked storm, we still forged ahead, hoping to get through the worst of it. I'll never forget passing the Montana Highway Department's wind sock that had been so whipped by the gales that the only thing left was the metal ring. *After that,* we passed a sign that said, CAUTION. HIGH WIND AREA AHEAD! The 70-mph winds and Montana's icy roads finally forced us to pull over in a truck stop, where we shared the crowded parking lot with countless semis and passenger cars. Although keeping Zinger in the trailer protected her from the wind, eventually I had to unload her and take her for a walk to loosen her up, as she had been standing for many hours already.

It was everything you don't want for a trailering trip: an icy parking lot, winds trying to rip off the trailer doors when you open them, and a crowded scene with many diesel engines rumbling. In spite of all this, Zinger was steadfast and reliable in her unloading, and when it came time to put her back in the trailer, she walked right in.

This is just one of Zinger's early road-trip stories. She followed me to Iowa, Illinois, and on to Colorado. Over the years, she and I have trailered to many lessons, shows, clinics, trail rides, and cattle drives. She's reliable. She's a veteran.

That's why I chose Zinger for the cover and for several of the training and traveling demonstrations inside the book. A horse like Zinger is worth her weight in gold. You can make your own horse a "Zinger" if you follow the advice and procedures you'll find here.

Happy trails and safe traveling to you.

PART I

THE TRAILER AND THE TRUCK

Your rig consists of a towing vehicle (usually a truck) and a trailer. The selection, use, and maintenance of your truck, trailer, and hitch are of major importance. To ensure safety for you and your horse and to protect the big investment you have in your rig, learn all you can about truck and trailer features, maintenance, and maneuvering.

An ideal horse-hauling rig is a ¾-ton or larger truck with dual rear wheels and a gooseneck trailer (it attaches to the pickup bed just ahead of the rear axle). This will afford the safest and most reliable hauling experience for you and your horse. I realize that most horse owners must make some compromises, though, so I would like to provide you with the information to help you make wise and safe decisions. If you already own a towing vehicle and must buy an appropriate trailer, *or* if you have a trailer and need to find a suitable towing vehicle, *or* if you are looking to buy both a truck and a trailer, you should become very familiar with the material in this section to help you make the right decisions.

Read about truck features, towing capacity, and maintenance in chapter 1. See the wide array of trailer designs, features, and options in chapters 2 and 3. Find out in chapter 4 why trailer maintenance is the best insurance policy for you and your horse. Learn how to maneuver your trailer and drive safely in chapter 5.

1

✦ THE TOWING VEHICLE ✦

There are many factors to consider when determining whether a vehicle is suitable for pulling a horse trailer. Your choice of vehicle will depend on the other uses you plan for it (commuter, family van), the number of passengers and type of cargo it will carry, the climate and terrain it will be used in, the extent of use for hauling, and the types of roads it will be used on (flat interstate, mountainous gravel).

Be sure your truck has the adequate weight, length, horsepower, and towing capacity to safely pull your trailer! Overloading a tow vehicle is not only illegal and unsafe but it will also take longer to stop the load, require more power to start the load moving, and cause increased wear and tear on the suspension, brakes, engine, and drive train.

Manufacturers rate each vehicle according to its towing capacity, but manufacturers' ratings are usually based on hauling a static load such as a boat and trailer. The demands on a towing vehicle are greater for live weight than for fixed cargo. When you are hauling horses, plan to haul about 25 percent less than the maximum load rating. It's better to have more truck and not need it than to need more truck and not have it.

Towing Capacity

Towing capacity is determined by many factors, including vehicle width, engine size, wheelbase, transmission, gear ratio, and the vehicle weight.

Vehicle Width

The width of your vehicle will greatly affect the stability of the rig when you are moving. Trucks with single rear wheels are adequate for two-horse trailers, but for larger trailers a truck with a wider rear end can increase rig stability by decreasing sway. Sway is often caused by the movement of live cargo. When a tow vehicle sways, the tire tread stays in place on the road but the sidewalls of the tires are pushed off to one side by the weight of the trailer. A truck with dual rear wheels (a "dually") not only increases vehicle width but also adds a second set of heavy-duty tires to the rear axle. The extra tires with heavy-duty sidewalls help counteract the tendency to sway, which results in increased stability and traction.

Engine

The towing vehicle's engine should have enough horsepower to haul the extra weight of a loaded trailer. The example truck (see chart on page 6 for metric conversions) has a 5.9-liter V-8 (360 cc) engine with 235 horsepower at 4,000 rpm, which is suitable for pulling a two-horse trailer. If you opt for a smaller engine, what you save in gas mileage may result in greater repair costs due to excess strain on the engine. In the mountains, figure your engine will lose 2 to 4 percent of its power for each 1,000-foot (305-m) increase in altitude.

If you plan to own your truck for a long time and will be doing a lot of long-distance hauling, especially in hilly or mountainous terrain, you might want to consider a diesel engine. Diesels are noisy, smelly, and higher priced; and they have higher repair costs and run on higher priced fuel than do gas engines. However, diesel engines have a longer life than gas engines and are very reliable for long-distance hauling.

Wheelbase

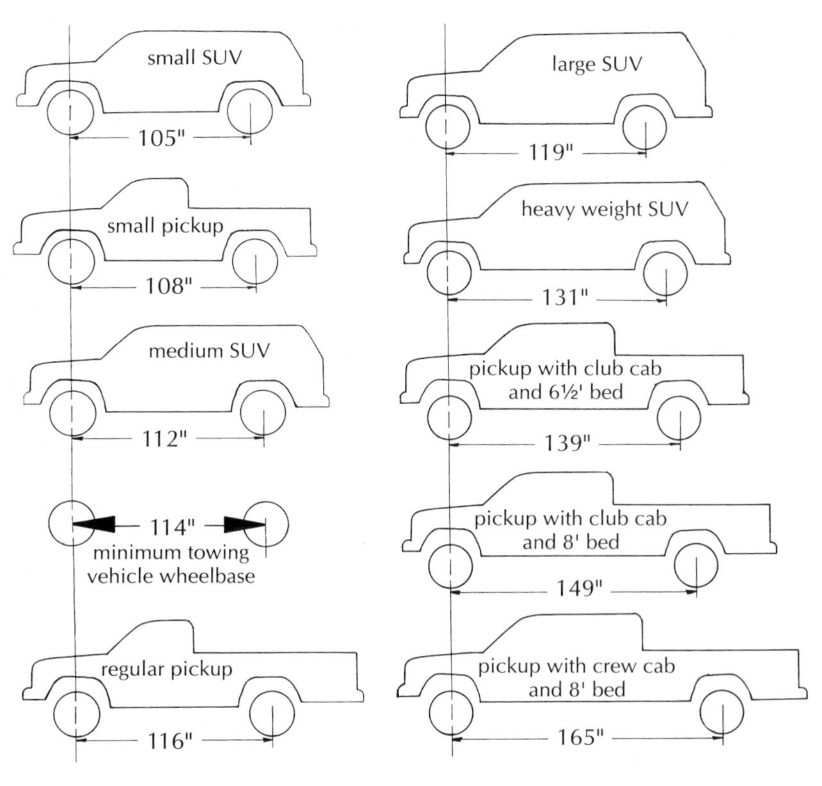

The wheelbase is the distance between the centers of the front and rear axles. The longer the wheelbase, the greater the towing stability when the rig is in motion. For a short, two-horse, straight-pull trailer (this attaches directly to the rear of the truck), the wheelbase of the towing vehicle should be an absolute minimum of 114 inches (290 cm). This eliminates small pickups and most sport utility vehicles (SUVs). To haul a two-horse trailer with a dressing room, you will need a wheelbase of at least 120 inches (305 cm); for a three- or four-horse trailer, look for a vehicle with a wheelbase of 139 inches (353 cm) or longer. The wheelbase is determined in large part by the length of your truck's cab and bed. Cabs are available in (from shortest to longest) standard, Club Cab (also called extended, super, and extra cab), and Crew Cab (also called four-door). Beds come in short (6½ feet [2 m]) or standard (8 feet [2.5 m]). Be aware that the longer the overall truck, the harder it will be to park and maneuver as an independent vehicle, an important consideration when your truck also serves as your town car. The wheelbase of the full size ½-ton pickup with a club cab and short bed in photos 1.3, 1.4, 1.5, and 1.14 is 139 inches (353 cm).

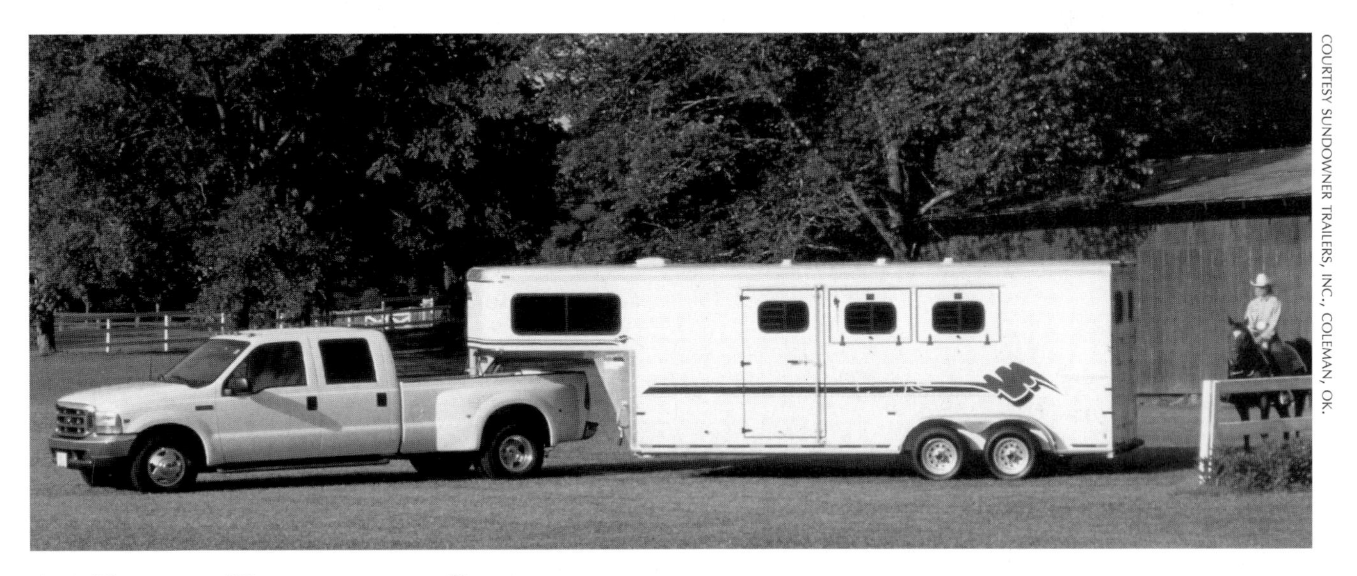

COURTESY SUNDOWNER TRAILERS, INC., COLEMAN, OK.

1.1 Dually Truck with Gooseneck

This is an ideal setup for hauling horses. This dually truck with crew cab has a more than adequate wheelbase. The three-horse, aluminum, slant-load, gooseneck trailer with dressing room is ready to take you and your friends safely to the next show or trail ride. It has the size and power to be a stable hauling vehicle, and the trailer is well suited for the truck.

Differential/Axle Ratio

The gear ratio of the differential or "rear end" (the gearbox on the rear axle) determines how the engine's power is transferred to the wheels. It is a measurement of how many times the drive shaft must turn for one rotation of the vehicle's wheels. The higher the gear ratio, the better the torque or pulling capacity, especially for starting, accelerating, and passing. Finding the ideal gear ratio is a compromise, however, because as the gear ratio goes up, the top end speed and the gas mileage go down. For occasional hauling, the 3.55:1 gear ratio in the example truck is about optimum for a ½-ton multipurpose vehicle. If you do a lot of hauling or pull a four-horse trailer or larger, look for a greater gear ratio. For ¾-ton trucks and heavier, a gear ratio over 4:1 is more suitable for hauling.

Transmission

Automatic transmissions are easier to drive, last longer, generally make for a more comfortable ride for the horses, and are usually rated higher than standard transmissions for their towing capabilities. A standard transmission (manual shift) can be tricky to master but it should get slightly better gas mileage than an automatic. However, when you're hauling a horse trailer, no matter which transmission you have, when it comes to gas mileage, you really don't want to know! If you choose an automatic, be sure you can lock out, or shift out of, overdrive (fourth gear). On some roads and with certain loads, the constant shifting in and out of overdrive can be annoying and cause unnecessary wear on the transmission.

Heavy-Duty Towing Package

A heavy-duty trailer-towing package is necessary for all loads over 2,000 pounds (907 kg), which means *all* horse trailers. The towing package should include a heavy-duty hitch (see chapter 2) and receiver, a heavy-duty radiator, a heavy-duty transmission with auxiliary oil cooler, heavy-duty suspension (springs and shocks), a heavy-duty battery and alternator, heavy-duty flashers, and a wiring harness. The wiring harness usually consists of seven wires connected to a plug, in order to accommodate electric brakes and trailer lights.

Look in the owner's manual for specifications on your vehicle or ask your dealer to help you determine whether you need to beef up your truck.

Two-Wheel Drive (2WD) vs. Four-Wheel Drive (4WD)

Those who live in snowy country or are going to be doing off-highway driving, such as is necessary at many trail ride sites, should consider 4WD. Four-wheel drive adds quite a bit of weight to the towing vehicle's Gross Vehicle Weight (GVW; see box), which adds stability to the towing vehicle but decreases its towing capacity and lowers gas mileage. A four-wheel-drive truck sits higher off the ground than does a 2WD, so it may require adjustments to the truck's suspension or hitch for the trailer to ride level.

Weight

Your towing vehicle must be heavy enough to control the load it is pulling. You will need to gather and calculate some numbers on your vehicles. Refer to the chart on page 6 for the stats on the example truck and trailers. Make similar worksheets for your own truck and trailer or those you are considering buying. More detailed information on these weights is given throughout this chapter.

Curb Weight (CW) is the weight of the empty vehicle. The best way to determine actual curb weight is to drive your empty truck onto a commercial scale. Actual curb weight usually includes a driver, a full tank of gas, fluids, and standard equipment such as spare tire and jack. To calculate an estimate of your truck's curb weight, first get the manufacturer's standard curb weight figure for your

TIP

You can find commercial scales that are long enough to weigh semi tractor/trailers at your feed mill, gravel pit, or at highway truck checks. Plan to pay a nominal fee such as $5 for the weight ticket.

Weighty Definitions

Curb Weight (CW). The weight of an empty truck plus gas and driver, *or* the weight of an empty trailer not hitched to a vehicle.

Payload. The maximum allowable weight that a truck is designed to carry, per manufacturer.

Gross Vehicle Weight Rating (GVWR). The maximum allowable weight of the truck plus its cargo, per manufacturer, *or* the maximum allowable weight of the trailer plus its cargo, per manufacturer (sometimes called the Loaded-Trailer Weight Rating, or LTWR).

Gross Vehicle Weight (GVW). The actual weight of the truck including all cargo, passengers, and the trailer tongue weight, *or* the actual weight of the trailer, including weight of the horses, tack, feed, water, and other cargo minus the tongue weight (sometimes called the Loaded-Trailer Weight, or LTW).

Gross Combined Weight Rating (GCWR). The maximum allowable total weight (per truck manufacturer) of the loaded truck (GVWR) plus the loaded trailer (GVWR).

Gross Combined Weight (GCW). The actual weight (GVW) of the loaded truck plus the actual weight (GVW) of the loaded trailer.

Tongue weight. The amount of the trailer's weight that is being transferred to the truck via the tongue or gooseneck.

Maximum Trailer Weight Rating (MTWR). The maximum allowable towing capacity of a vehicle (sometimes called the maximum tow weight rating); some manufacturers will also provide a maximum tongue weight rating, which is approximately 10 percent of the MTWR.

truck. Most truck dealers can provide you with the manufacturer's curb weight (with plus or minus 3 percent accuracy). You'll need to know the year, model number, engine, transmission, cab style, whether 2WD or 4WD, and other information to get the most accurate estimate from the dealer. (You can also get vehicle statistics from http://www.consumerguide.com.) The manufacturer's curb weight figure represents a vehicle with standard equipment. For the example truck, the dealer provided the figure of 5,000 pounds.

To find the curb weight of your truck from the manufacturer's figure, add the weight of nonstandard equipment you have installed on your truck such as a camper shell, running boards, or tonneau cover. This total is an estimate of your truck's actual curb weight. Your truck's *actual* curb weight is the weight on a commercial scale. The example truck's actual curb weight by scale is 5,300 pounds, which reflects 300 pounds (136 kg) of add-ons.

Gross Vehicle Weight Rating (GVWR) is the maximum allowable weight of the truck and its payload as established by the manufacturer. For safety, the towing vehicle should never weigh more than the GVWR.

1.2 The GVWR is listed on the Vehicle Identification Plate, usually located on the driver's door-latch pillar. The GVWR also appears on the window sticker of a new vehicle and in other paperwork and brochures. The GVWR of the example truck is 6,400 pounds.

Weight Chart and Worksheet

Example Truck (½-ton pickup)	Pounds (Kg)	Your Truck's/Trailer's Weight in Pounds (Kg)
Curb weight, per manufacturer	5,000 (2,268)	______
Payload, per manufacturer	1,400 (635)	______
GVWR, per manufacturer	6,400 (2,903)	______
GVW w/add-ons, by scale	5,300 (2,404)	______
GVW w/empty steel trailer attached, by scale	5,900 (2,676)	______
GVW w/empty aluminum trailer attached, by scale	5,630 (2,554)	______
GVW w/loaded steel trailer attached, by scale	6,260 (2,840)	______
GVW w/loaded aluminum trailer attached, by scale	5,900 (2,676)	______
MTWR, per manufacturer	7,300 (3,311)	______
GCWR, per manufacturer	12,500 (5,670)	______
GCW w/loaded steel trailer	11,300 (5,126)	______
GCW w/loaded aluminum trailer	9,940 (4,509)	______

Example Trailer A: two-horse slant-load steel with dressing room		
Curb weight, per manufacturer	3,600 (1,633)	______
Curb weight with options, by scale	3,860 (1,751)	______
Tongue weight, per manufacturer	540 (245)	______
Tongue weight percentage, per manufacturer	15%	______
Tongue weight, empty, by scale	600 (272)	______
Tongue weight percentage, actual	16%	______
Tongue weight, loaded	960 (436)	______
GVW (3,860-pound [1,751-kg] trailer, 2,000 pounds [907 kg] of horses, 140 pounds [64 kg] of tack, feed, water)	6,000 (2,722)	______
GVWR, per manufacturer	7,000 (3,175)	______

Example Trailer B: two-horse straight-load aluminum with dressing room		
Curb weight, per manufacturer	2,500 (1,134)	______
Tongue weight, empty, per manufacturer	300 (136)	______
Tongue weight percentage, per manufacturer	12%	______
Tongue weight, empty, by scale	330 (150)	______
Tongue weight percentage, actual	13%	______
Tongue weight, loaded	600 (272)	______
GVW (2,500-pound [1,134-kg] trailer, 2,000 pounds [907 kg] of horses, 140 pounds [64 kg] of tack, feed, water)	4,640 (2,105)	______
GVWR, per manufacturer	7,000 (3,175)	______

Payload is the maximum allowable weight of cargo, passengers, and tongue weight rating that the truck is designed to carry. This figure is available from the vehicle manufacturer or dealer.

GVWR – CW = payload
6400 – 5000 = 1400 for the example truck

Therefore, if you can't find the standard curb weight of your vehicle (and don't have access to a scale), you can still calculate it if you know the GVWR and payload.

GVWR – payload = standard CW
6400 – 1400 = 5000 for the example truck

Note: If you are trying to end up with an estimate of the *actual* curb weight using this method, you still must add the weight of nonstandard equipment that you have added to the truck.

Gross Vehicle Weight (GVW) is the actual weight of your loaded truck. It is the curb weight plus the weight of any passengers (other than the driver, who was already figured in the curb weight) and the weight of any cargo in the truck. The GVW should never exceed the GVWR. If you are pulling a trailer, the GVW will also include the tongue weight of the trailer (see later in this section, as well as chapter 2). The GVW is best determined by weighing, but it also can be estimated by calculation. The result should be compared to the GVWR. For the example truck, note that the GVW is 5,900 pounds with an empty steel trailer attached. Because the GVWR for this vehicle is 6,400 pounds, when two horses are in the trailer, the GVW of the truck increases to 6,260 pounds due to the increase in tongue weight. This "works" but is quite close to the 6,400-pound GVWR of the truck. It doesn't leave much leeway for passengers or cargo. (See other truck-trailer options later in this chapter.)

Gross Combined Weight Rating (GCWR) is the maximum allowable total weight of the entire rig: loaded vehicle plus loaded trailer. Get this figure from the truck's manufacturer. You can use the GCWR rating to help you calculate how much weight your truck is able to carry and tow. The GCWR of the example truck is 12,500 pounds.

1.3 The most accurate determination of a tow vehicle's GVW is obtained by weighing the loaded truck, with loaded trailer attached, on a commercial scale. Only the truck's wheels should be on the scale for this measurement. The trailer tongue weight will be a part of the GVW of the truck. The GVW of the example truck with Trailer A is 6,260 pounds; with Trailer B it is 5,900 pounds.

1.4 To obtain the GCW of your loaded rig, make sure that all tires of both the truck and the trailer are on the scale.

Gross Combined Weight (GCW) is the GVW (actual weight) of the fully loaded vehicle plus the GVW (actual weight) of the fully loaded trailer. You can calculate an estimate of the GCW, but the most accurate determination is to weigh your loaded truck and trailer on a commercial scale. The GCW of the example truck with loaded Trailer A is 11,300 pounds; with loaded Trailer B, it's 9,940 pounds. The GCW should never exceed the GCWR.

Calculate your rig's GCW, and compare it to the GCWR.

CHECKING GCW

Item	Example	Your rig's weight in lbs (kg)
GCWR	12,500 (5,670)	____________
GVW truck	5,300 (2,404)	____________
GVW Trailer A	6,000 (2,722)	____________
GVW Trailer B	4,640 (2,105)	____________
GCW with Trailer A	11,300 (5,126)	____________
GCW with Trailer B	9,940 (4,509)	____________

Following are discussions of other safe and unsafe truck-trailer combinations.

Combination 1. Use a ¾-ton truck to pull the steel trailer with a dressing room from the previous examples. This would be the best option because of the GVW and tongue weight of this trailer. A ¾-ton truck's GVWR will be higher than that of the ½-ton truck, so the tongue weight should pose no problem. Also, the GCWR of a ¾-ton will be higher than that of the ½-ton, so you will have a greater margin of safety and more cargo capacity.

Combination 2. Pull a lighter trailer with the ½-ton truck. You could choose a two-horse steel trailer without a dressing room (2,500- to 3,000-pound [1,134- to 1,361-kg] curb weight) or a two-horse aluminum trailer with a dressing room (2,500-pound), such as Trailer B in the examples and shown in photo 1.5. Either of these choices has a lower trailer GVW and lower tongue weight, which keeps the GVW of the truck and the GCW safely below the ratings.

Combination 3. Using a small to medium sport utility vehicle (SUV) or small pickup for towing a horse trailer is not a safe option for several reasons. Most small SUVs and small pickups have a tow rating well under 5,000 pounds (2,268 kg). And for some of those vehicles (and with good reason), you can't buy a hitch rated high enough to pull 5,000 pounds.

1.5 This two-horse aluminum trailer with dressing room has a GVW and tongue weight suitable for this ½-ton truck; a ¾-ton truck would be even better.

Even a medium SUV, such as a Ford Explorer, isn't a good choice for hauling a horse trailer. First of all, the Explorer's wheelbase is less than 112 inches (285 cm), which is below the absolute minimum of 114 inches (290 cm) for safety. If it has an automatic transmission, a heavy-duty towing package, and a high-performance axle, the towing capacity is 5,700 pounds (2,586 kg), which means it "technically" would be capable of pulling a light two-horse trailer (no dressing room) with two average-size horses. However, the curb weight of the vehicle is 4,500 pounds (2,041 kg) and the GVWR is 5,000 pounds (2,268 kg), which leaves only 500 pounds (227 kg) for tongue weight, passengers, and cargo. Even though the GCWR is about 10,000 pounds (4,536 kg), just about any two-horse trailer loaded would transfer too much tongue weight for the vehicle's GVWR, and the combined weights would be higher than the GCWR.

An alternative, if your heart is set on a SUV, is to use a larger one, such as the Ford Expedition. It has a 119-inch (302-cm) wheelbase, and if outfitted with automatic transmission, a V-8 engine, and proper towing package, it would have a 12,500-pound (5,670 kg) GCWR and a trailer-towing capacity of 7,000 pounds (3,175 kg). Especially if you use a weight-distribution hitch (see photo 1.10), it could work for a two-horse trailer.

General Guidelines for Safe Vehicle/Trailer Combinations

- Full size ½-ton truck for a two-horse (no dressing room) straight-pull trailer that weighs up to 6,000 pounds (2,722 kg) when fully loaded.
- ¾-ton dually (four rear wheels on the truck) for two-horse straight-pull with dressing room or three- to four-horse gooseneck trailer that weighs up to 7,000 pounds (3,175 kg) when fully loaded.
- 1 ton dually for five- to six-horse gooseneck trailer that weighs up to 10,000 pounds (4,536 kg) when fully loaded.

The designations ½-ton, ¾-ton, and 1-ton indicate a vehicle's approximate load-carrying capabilities. As you go up in numbers, it means the vehicle is capable of carrying more weight, is more heavy duty, and has higher clearance but has a stiffer ride and gets lower gas mileage.

Trailer Tongue Weight

A digression from truck talk is necessary to determine trailer tongue weight. Tongue weight refers to the amount of the trailer's weight that is being transferred from the trailer to the truck.

While a straight-pull rig transfers tongue weight to only the rear axle of the towing vehicle, a properly mounted gooseneck attachment results in a portion of the tongue weight of the trailer being distributed to the front wheels of the towing vehicle as well. Straight-pull trailers usually transfer 10 to 15 percent of their weight to the towing vehicle as tongue weight. Gooseneck trailers transfer approximately 25 percent of their weight to the towing vehicle as tongue weight.

Too much tongue weight can cause suspension and/or drive-train damage. With a straight-pull, too much tongue weight may press down the rear end of the towing vehicle, resulting in the front wheels lifting up to the point where steering and braking are seriously affected. A trailer that is heavily loaded in the front or is unbalanced could deliver more tongue weight to the truck than is desirable.

Determining Tongue Weight

You can determine tongue weight in three ways:

1. Ask the dealer. The figure he gives you will be an estimate for a standard trailer. The dealer figures supplied for both of the sample trailers were somewhat lower than the weight obtained from commercial scales. This discrepancy could be due to options added by the manufacturer (sidewall mats, spare tire, saddle racks, for example) which will affect overall weight and tongue weight, especially if the add-ons are in front of the trailer's axles. The difference between manufacturer figures and actual scale weight could also be a scale-calibration issue, making up a difference of up to 100 pounds (45 kg) measured on scales designed for semi trucks that weigh many tons.

2. Weigh your truck, then weigh your truck with the trailer attached. First drive your truck alone onto a scale. The example truck (see page 6 for metric conversions) weighs 5,300 pounds. Then attach your trailer. Make sure that the tongue, coupler, and receiver are horizontal and level when the trailer is fully hitched to the truck. Drive onto a scale, but be sure that only the truck's tires are on the scale. The example truck with Trailer A weighed 5,900 pounds. To determine the tongue weight, subtract the truck's weight (GVW) from the GVW with trailer attached. In the example, 5900 – 5300 = 600 pounds for the tongue weight of the empty steel trailer. For Trailer B, 5630 – 5300 = 330 pounds. You can use this method with an empty or a loaded trailer. But if you do this for a loaded trailer, you will need to make two trips to the scale because you will have to weigh your truck separately (without trailer attached) to find its actual GVW, and you should never unhitch a loaded trailer.

3. Weigh your truck with the trailer attached, then unhitch the trailer. You can also calculate the tongue weight for an empty trailer by first taking the GVW of the truck with trailer attached. Then, with the truck still on the scale, block the empty trailer securely, unhitch it, and make sure that the coupler is not touching the ball of the truck's receiver and the trailer jack is not on the scale. Note the weight of your towing vehicle now, making sure all passengers and cargo have remained in the vehicle for both measurements (5,300 pounds for the example vehicle). Subtract the second figure from the first figure: For Trailer A, 5900 – 5300 = 600 pounds tongue weight. For Trailer B, 5630 – 5300 = 330 pounds. This is the tongue weight of the empty trailer. Never unhitch a trailer with horses in it. This method is to be used with an empty trailer only.

Too little tongue weight can actually lift the rear end of the towing vehicle up by the ball, which reduces rear-wheel traction. When a trailer is loaded heavily at the rear, the tongue may elevate. This type of imbalance is most likely to cause dangerous trailer sway and fishtailing. The truck is no longer in control because the heavy trailer is like "the tail wagging the dog."

Tongue weight percentage. To determine what percentage of the trailer's weight is being transferred to the truck via the tongue, divide the tongue weight by the trailer weight. For Trailer A, the tongue weight (600 pounds) divided by the trailer weight (3,860 pounds) results in a tongue weight of 16 percent. For Trailer B, the tongue weight (330 pounds) divided by the trailer weight (2,500 pounds) is 13 percent.

Theoretically, the tongue weight percentage should be approximately the same whether or not the trailer is loaded, as long as the load is fairly evenly balanced. So once you have calculated the empty trailer tongue weight and percentage, you can use that percentage to calculate an estimate of the loaded trailer tongue weight. For Trailer A, if the loaded trailer weight is 6,000 pounds (3,860-pound trailer, 2,000 pounds of horses, 140 pounds of tack, water, and feed), then 16 percent is 960 pounds, the weight transferred to the truck. When added to the example truck's GVW of 5,300 pounds, the resulting 6,260 pounds is close to the vehicle's GVWR of 6,400 pounds.

Trailer B's loaded trailer tongue weight, 13 percent of 4,640, is 600 pounds. This makes the GVW of the truck 5300 + 600 = 5900 pounds, 500 pounds under the GVWR.

Give and take. When your trailer is hitched to the truck, add the tongue weight to the truck's GVW and subtract it from the trailer's GVW. In essence, the trailer is giving that weight to the truck to carry. Thus, when Trailer A is attached, the truck's GVW is 6,260 pounds and the truck is towing 5,040 pounds (2,286 kg; 6000 – 960 = 5040 pounds). When Trailer B is attached, the truck's GVW is 5,900 pounds, and the truck is towing 4,040 pounds (1,833 kg; 4640 – 600 = 4040 pounds).

Rules of Thumb

The following rules of thumb will help you to estimate safe towing capacity and to get an idea of the relationship between the towing vehicle and the loaded trailer. However, always weigh and calculate your rig's components to be certain and safe.

1. Usually a towing vehicle can safely pull approximately 1⅓ times its curb weight.

For the example truck with a manufacturer's curb weight of 5,000 pounds, 1⅓ times is 6,650 pounds (3,016 kg). The MTWR for this vehicle is 7,300 pounds. Because it is sometimes difficult to obtain a MTWR from a manufacturer, our rule gives a safe estimate.

2. Especially for interstate driving, the GVW of the truck should be at least 75 percent the GVW of the trailer.

For example, the GVW of Trailer A is 6,000 pounds; of Trailer B, it is 4,640 pounds. The towing vehicle should weigh at least 75 percent of 6,000 pounds, which is 4,500 pounds (2,041 kg), to haul Trailer A and at least 75 percent of 4,640 pounds, which is 3,480 pounds (1,579 kg) to haul Trailer B. With a GVW of 5,300 pounds, this truck has adequate weight.

Tires

Your truck's tires should be of the appropriate size for the vehicle and inflated with equal, optimum pressure. Tires should be in good condition. If you are hauling a two- or three-horse trailer, use six-ply rating, Load Range C radial tires with at least 15-inch (38-cm) rims. For four-horse trailers, use eight-ply rating, Load Range D, and 16-inch (41-cm) rims.

Ply rating is the strength index of a tire. Plies are the layers of parallel cords coated in rubber that form the body of a tire. The actual number of plies used to be marked on the sidewall of the tire, but this system is being replaced by the term *Load Range*, which is a range of letters A through E, where A begins with two-ply rating and E is the highest rating. On tires with lower-ply ratings, the sidewalls tend to flex excessively under weight, which can result in sway.

Tires with an open, "aggressive" tread design, such as all-terrain tires, have better traction in mud and snow. Tires with a smoother tread, such as highway and all-season tires, are quieter, run cooler, and promote better gas mileage. With four-wheel drive, aggressive tread designs usually aren't necessary.

Truck's Hitch

The hitch consists of the truck's hitch and the trailer's hitch (see page 19). The truck's hitch consists of the receiver, the ball mount, and the ball. The receiver (framework) is bolted to the truck's chassis. The ball mount (often removable) is attached to the receiver. The ball is attached to the ball mount. The trailer coupler connects to the ball. Each component must be rated for the GVW of the trailer and the tongue weight of the trailer. There should be a stamp or decal on each item listing its rating. Generally, the tongue-weight rating is 10 percent of the hitch rating. A hitch rated for 10,000 pounds (4,536 kg) likely has a 1,000-pound (454 kg) tongue-weight rating.

CAUTION

If you order a heavy-duty towing package on your new truck, there's a good chance the manufacturer or dealer will install a 5,000/10,000 hitch and tell you that it's good to tow 5,000 to 10,000 pounds (2,268 to 4,536 kg). That is not the complete truth!

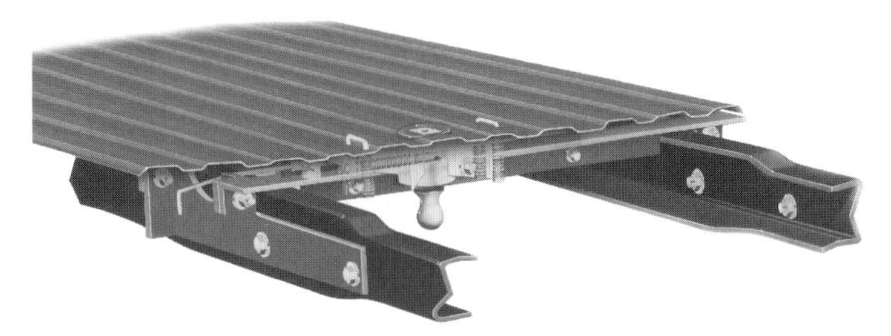

1.7 Turnover Gooseneck Ball
A turnover gooseneck ball has a unique design that allows you to turn it upside down and store it under the pickup bed when not in use. (Courtesy B&W Custom Truck Beds, Inc., Humboldt, KS.)

1.6 Mud Flaps
Mud flaps will protect your trailer from gravel, mud, and dirt thrown by the rear tires of the truck.

1.8 Straight-Pull Ball Mount
The ball mount for a straight-pull trailer is a unit that slides into the square tube *receiver* of the hitch and is usually fastened in place with a hitch pin and clip. The ball is bolted onto the ball mount. The drop height (the amount the ball is lower than the receiver) of ball mounts varies from 1½ to 8 inches (4 to 20 cm) to accommodate the various heights of truck and trailer so that the trailer is level. Ball mounts can also be turned over and the ball bolted to the other side if the trailer's coupler is higher than the hitch. Remove ball mounts when not in use, as they are easily stolen and a common cause of bruised shins.

1.9 Weight Ratings Label

Because there is no standardization among manufacturers as to what constitutes a Class III or a Class IV hitch, do not use class designation when choosing or ordering a hitch. Instead, use actual weight ratings, and know what they mean.

> **WARNING**
>
> When you order a new truck with a heavy-duty towing package, don't assume that the hitch the manufacturer puts on the truck will be adequate for the truck's MTWR. Some manufacturers do not offer a factory hitch that is adequate for the MTWR of their trucks. You might have to buy an after-market, heavy-duty hitch if you want to pull a loaded two-horse trailer without using a weight-distribution hitch. Be sure that the lowest number of the hitch's maximum-trailer-weight rating is higher than the loaded trailer weight you plan to pull.

There will be two weights stamped or labeled on the frame portion of the hitch. The lower number refers to the hitch's capacity in a "weight-carrying" situation, with a customary ball mount, not a weight-distribution hitch. A 5,000-pound-maximum (2,268-kg) trailer-weight rating and 500-pound-maximum (227-kg) tongue weight is inadequate for most two-horse trailers, yet it is one of the most common hitches put on vehicles.

The higher figure of the rating indicates the capacity of the hitch when a weight-distribution hitch is added to the truck and trailer. With the hitch shown in photo 1.9, a truck can tow up to 10,000 pounds (4,536 kg) and carry up to a 1,000-pound (454-kg) tongue weight *without* the addition of a weight-distribution hitch.

For almost any straight-pull trailer, it's best to use at least a 10,000/12,000-pound (4,536/5,443-kg) hitch. Be sure the ball mount and ball are rated for those weights as well.

Weight-Distribution Hitch

A weight-distribution hitch is a substitute for or an add-on to a receiver hitch to help you increase your receiver's towing capacity up to 14,000 pounds (6,350 kg) and its tongue-weight capacity up to 1,400 pounds (635 kg). A weight-distribution hitch has two steel bars that transfer tongue weight to the front axle of the tow rig and to the trailer axles. The distribution results in increased stability for the whole rig. Although a certain amount of tongue weight is necessary for the truck to have control of the trailer, when the tongue weight is too heavy, weight is lifted from the truck's front axle and transferred to its rear axle, thus increasing the rear loading. For tongue weights over 700 pounds (318 kg), the truck can be unbalanced enough that control may be affected. If a truck is light in the front end, the steering, turning, and braking will be negatively affected and the truck will lose the benefit of all four of its tires to help reduce sway. You can't use a visual assessment to determine if tongue weight is appropriate because often the springs are strong enough that the truck looks fine (not raised in front or pulled down in back) even though the rig is not balanced.

A weight-distribution hitch (photo 1.10) has two V-shaped arms, called spring arms, that run parallel underneath the A-frame tongue of the trailer. Two cuffs are attached to the trailer frame, one on each side, in line with the ends of the arms. When you hook up the trailer, chains at the ends of the spring arms connect to the cuffs. On each cuff is a type of rocker lever that tightens the chains and basically lifts the truck and trailer at the hitch point, transferring weight to the front axle of the truck and the rear axle of the trailer. The trailer is now being supported or towed from three points (the ball and two cuffs), as opposed to just one point (the ball) in a normal hitch assembly.

1.10 WEIGHT DISTRIBUTION HITCH
The adjustable ball mount allows the hitch to be used with towing vehicles of various heights. Note the open hooks on the safety chains. These could easily jump off their points of connection.

Sway Controls

Sway controls may be an add-on or part of a weight-distribution hitch. They are designed to reduce sway, but before installing them, do be sure they're necessary.

Check that your tongue weight is in a normal range. If it's high, make sure you have an adequately rated receiver on your truck. You can also consider using a weight-distribution hitch. If the tongue weight is too low, the trailer will sway the truck, so increase the tongue weight by moving your trailer load forward. If you are using tires that are rated too low for your load, the weak sidewalls could be causing the sway. Low air pressure in either the truck or the trailer tires can also result in sway.

When the coupling is too high (see photo 1.15), the floor of the trailer will slant downward toward the rear of the trailer. Excess weight then will be riding on the rear axle of the trailer, which can actually lift up the rear of the towing vehicle and cause vehicle sway, as well as make the ride uncomfortable for the horses.

If the coupling is too low, the floor of the trailer will slant downward toward the front of the trailer. This will cause excess wear on the ball and excess tongue weight on your towing vehicle.

Once you have evaluated all of these causes of trailer sway, you may want to consider auxiliary anti-sway devices. There are basically two types of sway controls: a friction sway control and a cam sway control. Friction sway control uses friction to stiffen the coupling between the tow vehicle and the trailer by means of a two-piece rod, somewhat like a shock absorber, that mounts between the ball mount and the trailer A-frame. The degree of stiffening or friction is adjustable to suit various trailer weights and towing conditions. Friction sway controls don't prevent sway; they simply resist sway forces once they have started.

A cam sway control works to prevent sway in the first place. When towing in a straight line, cams on either side of the trailer are locked in position. This essentially creates a rigid connection between the tow vehicle and the trailer and minimizes sway caused by high crosswinds and passing vehicles. When cornering, the cams automatically slide out of their locked position to permit full-radius turns. Yet when the maneuver is short and abrupt, like a sudden swerve or a wheel dropping off the road, the cams lock to help the tow vehicle retain control.

1.11 Ball Rating

The ball must be rated for the GVW of the trailer and of the proper size to fit the trailer's coupler, or the trailer could bounce off the ball. The size and weight rating are stamped on the ball. The most common straight-pull ball size on older trailers was 2 inches (5 cm); today's straight-pulls and goosenecks most commonly use a $2\frac{5}{16}$-inch (6-cm) ball. Straight-pull balls should be rated for at least 10,000 pounds (4,536 kg). The one in this photo is rated to pull only 5,000 pounds (2,268 kg). Gooseneck receivers and balls should be rated for at least 25,000 pounds (11,340 kg). If the stamp is worn off, you can measure ball size with a pair of vernier calipers. The shaft of the ball should be the same size as the hole in the ball mount.

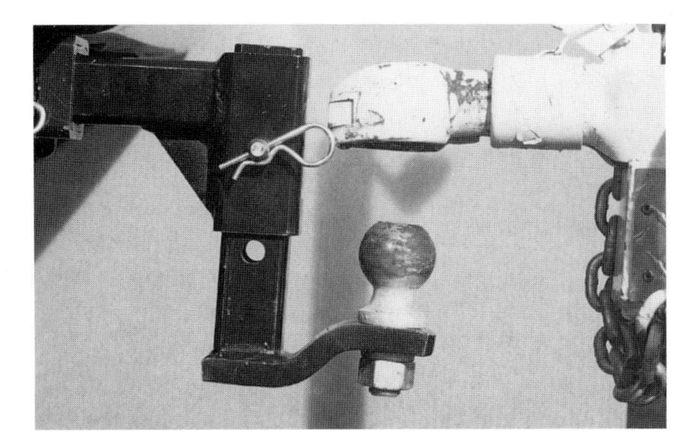

1.12 Adjustable Ball Mount

If you pull several types of trailers with your vehicle, consider an adjustable ball mount that can be used at a variety of heights. By moving the ball mount up and down or flipping it over, you are adjusting the height. Be sure the ball mount and ball are rated for the weight of the trailer you are pulling. (Also see photo 1.10 for another style of adjustable mount.)

1.13 Adjustable Gooseneck

Some gooseneck trailers have adjustments in the neck to allow you to raise or lower the trailer to fit trucks of various heights. To adjust the height of this gooseneck, loosen the bolts and move the tube up or down as needed. However, be aware that if you shorten the neck of a gooseneck to fit a tall truck, the trailer may come in contact with the tailgate or bed corners when turning. (See photo 2.7 for another way to make the gooseneck and truck bed a compatible height.)

1.14 On the Level

When the loaded trailer is attached to the truck, the trailer floor and the tongue should be level. Set the ball mount at the proper height so the trailer is level. The coupler should rest on the ball in optimum balance, so 85 to 90 percent of the trailer's weight is borne by its wheels.

1.15 A Rearing Trailer

This rig is not level. And, more importantly, this trailer is hitched to the truck's bumper, which likely is not rated high enough to pull a loaded two-horse steel trailer. A properly attached straight-pull receiver with a stepped-down ball mount would put this trailer at the correct level. With a straight-pull, always use a ball mount with the proper drop for the rig. With a gooseneck, either have the suspension adjusted to lower the truck or adjust the neck of the gooseneck coupling to be sure the rig is level.

Brakes

Power steering and power brakes are a must for a towing vehicle. Antilock brakes, which create a rapid pulsation instead of a brake lock when brakes are used hard, can prevent a skid. They should be at least on the rear wheels, and on all four if possible.

A towing vehicle's brake system is designed and rated for operation of the vehicle's GVWR, not the GCWR of the truck and trailer. Therefore, a separate brake system must be installed and used on all horse trailers. Every state requires trailer brakes for trailers weighing more than 3,000 pounds (1,361 kg). You can choose between electric and hydraulic trailer brakes (see chapter 3 for more on trailer brakes).

Servicing the Towing Vehicle

Routine servicing of the engine, cooling systems, suspension, tires, wheel bearings, brakes, and other mechanical components not only prolongs the life of a vehicle, but also allows you or your mechanic to uncover problems before they become an emergency.

1.16 ELECTRIC BRAKE CONTROLLER

This hand controller for electric trailer brakes has been mounted under the dashboard between the accelerator pedal and the 4WD-transfer case. This controller allows you to operate the trailer's brakes separately from the truck's. (See chapters 3 and 5 for more about electric brakes.)

1.17 SERVICE SCHEDULE

Follow the heavy-duty service schedule in your owner's manual. When towing or driving on gravel roads, a vehicle requires more frequent air filter and oil changes. Some common problems you may encounter when towing are an overheated engine or transmission, a flat tire, and brake or hitch failure.

2 THE TRAILER

Safety should be the number one concern when selecting a trailer for hauling horses. If you are considering a particular trailer, look at the design and features with your horse's comfort and well-being in mind. Put function before frills in all of your decision making. Insist on safety features such as rubber bumpers on door sills and double-lock door latches before considering the color of the carpeting in the tack room. Evaluate the interior dimensions, suspension, flooring, and ventilation to ensure your horse will have the most comfortable ride possible.

There are basically three styles of horse conveyances to choose from: the fully enclosed trailer, the stock trailer, and the horse van.

Most enclosed trailers are suitable for long trips. Enclosed trailers are available in one-, two-, three-, four-, and six-horse models. They range from a basic two-horse trailer with under-manger tack compartments to custom-designed luxury models with living quarters, tack, and feed rooms.

The open design of stock trailers can limit their use for long trips and during cold or wet weather. Stock trailers have slatted sides and economy interiors; thus, they weigh less and cost less than an equivalent-size enclosed trailer. However, because of the stock trailer's partially open sides, horses can get dusty, cold, and sometimes wet when riding in them. If you are buying a stock trailer, be sure that you choose one that is tall enough for horses. Many short stock trailers, perfectly suitable for cattle, sheep, or hogs, are much too low for horses. (See more on specific dimensions on pages 30–31.)

Vans are the most comfortable means of conveying a horse a long distance such as across the country. A van is essentially a stable on a truck. The truck and the horse compartment are combined into one vehicle. They can be 13 to 26 feet (4 to 8 m) long and may be designed for three to nine horses. Vans are generally more comfortable than a trailer for the horse. They have better suspension, which is less fatiguing to a horse's legs. They are heavier and better insulated than most trailers, which results in less road vibration and noise and contributes to a more stable temperature. Vans usually allow more head room for the horses, and the stalls often face each other so that horses can look at other horses rather than at a wall. Vans are the most expensive of the three types of haulers, and range from $40,000 to more than $100,000.

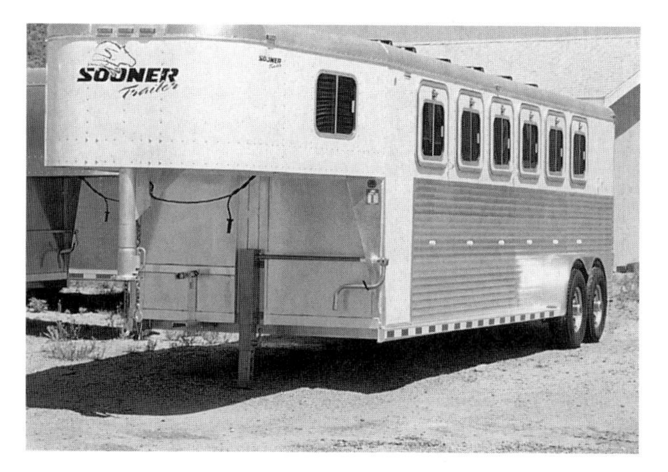

2.1 **Enclosed six-horse, slant-load, gooseneck trailer**

2.2 **Stock trailer**

2.3 **Horse van**

Trailer Hitches

There are three types of trailer hitches (couplers): the straight-pull, the gooseneck, and the fifth wheel. There are advantages and disadvantages to each of the systems. The type of hitch you choose will depend on the type of trailer you want to pull and the type of towing vehicle you have.

Straight-Pull Trailer

A straight-pull trailer (see photos 1.5 and 1.14), also known as a tagalong, is attached to the truck via a frame hitch. A straight-pull trailer is often incorrectly called a bumper-pull trailer (see the Caution box). A frame hitch is bolted directly to the frame of the towing vehicle. Frame hitches are rated according to the weight they can tow.

Note: Because there is no standardization among manufacturers as to what constitutes a Class III or a Class IV hitch, for example, do not use class designation when choosing or ordering a hitch. Instead, use actual weight ratings and know what they mean. (See chapter 1 for more information.)

Caution

Many straight-pull (tagalong) trailers are incorrectly called bumper-pull trailers, even by manufacturers. This stems from the days, not too long ago, when vehicles and their bumpers were made of very heavy materials and trailers were actually attached to the bumpers. Some full-size autos of the 1970s, with their long wheelbase, low center of gravity, heavy curb weight, and V-8 engines, were rated and able to tow substantial trailers attached to their bumpers. Not any more. Many of today's truck bumpers are very lightweight and could easily be damaged or come loose if a horse trailer were attached to them. Because most bumper hitches on light trucks and SUVs are rated for very light duty, under 3,500 pounds (1,588 kg), they are not safe enough to pull a loaded two-horse trailer.

So even if you buy a trailer that the brochure calls a "bumper pull," *never* attach a horse trailer to your vehicle's bumper. Even if a bumper is rated to pull the GVW of a trailer, it still is a bad idea. For one thing, the bumper would likely be too high, resulting in an unlevel trailer, and second, the trailer would follow so closely to the truck that it could hit the corner of the truck when turning. A frame hitch with ball mount locates the point of attachment of the trailer low enough to keep the trailer level and about 8 inches (20 cm) behind the bumper, which results in better clearance on tight corners.

Tell your trailer manufacturer and dealer that it is time for them to update their outmoded and confusing terminology.

Fifth Wheel or Gooseneck?

It is commonly thought that a fifth-wheel and a gooseneck hitch are the same. Though both attach to a point in the bed of the truck, the connections differ.

The fifth wheel used for large horse trailers is a small version of the type seen on semi trucks. Rather than a ball for attachment, there is an angled disk mounted on the truck's bed. If you are planning to haul six or more horses, you may want to consider a fifth wheel.

Far more common, though, is the gooseneck hitch. Only the ball is visible above the pickup's bed. Underneath the bed are supporting rails that are welded or bolted in place: These serve as the support structure for the hitch assembly. Some gooseneck balls are permanently mounted in an upright position in the bed of the truck, making it difficult to use the truck bed for certain types of hauling. Other gooseneck balls fold under the bed (see photo 1.7), or you can remove them entirely when not in use, leaving the bed flat.

Which is better, a straight-pull or gooseneck rig? If your towing vehicle is a pickup or a flatbed, you can choose between a frame hitch and a gooseneck hitch. If your towing vehicle is an SUV, has a camper, or includes a permanent tonneau cover, you will need to use a frame hitch along with a straight-pull trailer.

2.4 Gooseneck
The receiver for a gooseneck is attached near the center of the bed of the truck. The gooseneck coupler is at the end of the neck at the front of the trailer. Gooseneck rigs are recommended for any trailer that will haul more than three horses.

2.5 Straight-Pull
This straight-pull coupler and receiver shows a raised jack wheel, crossed safety chains, and level tongue. Safety chains on all trailers are required by law in most states.

2.6 Trailer Hitch Lock
Certain trailer locks can be used when the trailer is hitched to the truck or alone when the trailer is being parked or stored. This trailer lock slips over the coupler and requires a padlock.

2.7 Custom Bed Drop

Here a gooseneck trailer is hitched to the truck bed. Notice the custom drop in the bed to fit the height of the trailer. Note, too, the emergency trailer brake switch (black box), the safety chains, and the thick, black electrical cable (wiring harness) containing light and brake wires.

Additional Gooseneck Notes

With the popularity of 4WD short-bed, extended-cab trucks comes the question, "Can a gooseneck be used with a short-bed truck?" Sure, as long as you don't turn! Now for a more serious response: With some full-nosed gooseneck trailers, the trailer could hit the back of the cab when turning. You can compensate somewhat by installing the hitch a bit farther to the rear, but generally, you don't want the hitch to be over or behind the rear axle. Usually a gooseneck hitch is installed 2 to 6 inches (5 to 15 cm) in front of the rear axle. If you want a short bed, you'll have to choose your gooseneck trailer carefully.

Trucks have been getting taller, especially due to the popularity of 4WD. Many trailer manufacturers have made adjustments to their trailers to accommodate the higher beds, so their trailers will clear the bed rails and tailgates of taller trucks. The only way to determine whether a truck and trailer will match is to hook them up and see. (See photo 1.13 for an adjustable gooseneck.)

Pros and Cons of Straight-Pulls (Tagalongs) and Goosenecks

Trailer Type	Pros	Cons
Straight-pull	Most common in used trailer market; usually more adaptable between vehicles; most affordable — generally, 20% less expensive than gooseneck; good choice for two-horse trailer; tracks closer to vehicle than gooseneck, so don't have to take corners so wide.	Can be harder to maneuver; operator can't see ball when hooking up; puts more stress on rear end of vehicle; more prone to sway and fishtail.
Gooseneck	Better maneuverability in tight spots because of smaller turning radius; best choice overall and essential for four-horse trailer or more; distributes weight more evenly between both axles of towing vehicle; safest trailering method; nose area over gooseneck provides additional storage or sleeping space.	Requires open-bed vehicle with gooseneck receiver in bed; generally, costs 20% more than a straight-pull trailer; must take wider turns, as trailer tracks to the inside of truck on turns (cuts corners) and can run over curbs, etc.; requires a truck with greater towing capacity; less pickup bed storage area; must usually get into bed to hitch up trailer; more difficult to get an emergency tow; requires larger indoor storage area.

Floor Plans

The variety of trailer floor plans is almost limitless, especially when you consider that many manufacturers will build a horse trailer to your specifications. There is a lot of choice when it comes to selecting the number of stalls, the configuration of stalls, and the size and shape of the tack room, dressing room, and even living quarters that you'll find in a horse trailer. Each floor plan has advantages and drawbacks. (Some common configurations are shown in the following pages.)

Stock Trailer

2.8 Stock Trailer
The large, open spaces of stock trailers are quite versatile. You can tie horses in an open trailer or they can be hauled loose. There is usually a hinged center divider that can separate the front from the back or can be fastened along the wall as shown here. Often, hauling loose works well for a mare and her foal. Some horses exhibit less stress when they ride in a stock trailer with slatted sides (compared to a fully enclosed trailer) because they can see out, move around, and find a comfortable traveling stance. Research has shown that when given the choice, many horses like to ride facing the back of the trailer at an angle. However, if more than one horse is being hauled in a stock trailer, whether they are loose or tied, they must get along very well or they could injure each other. Also, loose horses may contribute to trailer sway. With open-floor trailers, use hay bags for feeding.

Mats should fit snugly to stay in place or be fastened down. The narrow strip mats in this open-style trailer could easily become dislodged by a scrambling horse and get tangled up in the horse's legs.

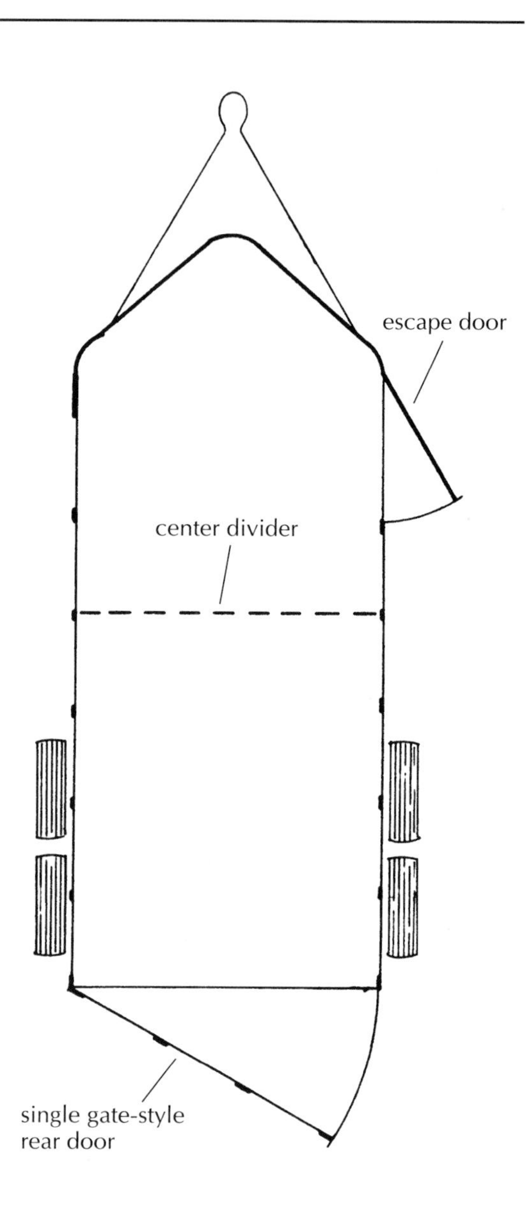

Straight-Pull Stock Trailer with one large rear door, center divider, and one escape door. Slatted sides instead of windows.

In-Line Trailer

2.9 IN-LINE TRAILER

In-line trailers are available in one- and two-horse models. This is a one-horse in-line trailer. In a two-horse in-line, the horses travel facing forward and are in single stalls one in front of the other. The second horse is loaded behind the first horse, head to tail. The second horse will not have a manger. Unlike tandem axles, which are close together, the axles of an in-line trailer are spread far apart, like a wagon, making this a stable trailer with a very light tongue weight. However, when backing, the trailer pivots at the hitch as well as at the front axle, which makes it maneuver more like a hay wagon. Consequently, in-line trailers can be very difficult to back — like pushing a rope down the street. Because the trailer is only one horse wide, it is narrow and thus stays on the road better than does a wider trailer. It is thought to be one of the safest designs in the case of an accident, because the horses don't get scrambled together and can be removed either through the roof or via the back door. Although no longer in production, they are available used.

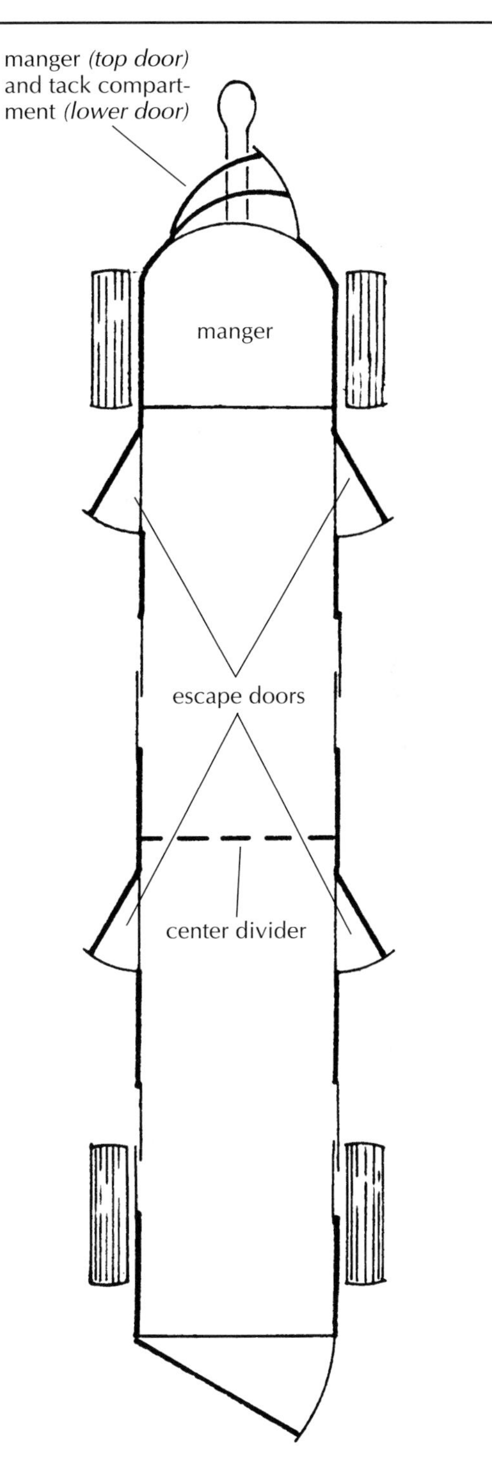

STRAIGHT-PULL, TWO-HORSE IN-LINE TRAILER with one rear door, one manger door, and one tack compartment in front, four escape doors, and four sliding windows.

Straight-Load Trailers

2.10 Straight-Load Trailer

Straight-load (side-by-side) trailers have been the most common two-horse trailers for many years. This old steel trailer has a minimal center divider, which lets a horse step sideways to balance. The stalls are too narrow and short for many of today's taller horses. Some four- and six-horse side-by-side trailers are configured so that some of the horses face forward and some face to the rear. Straight-load trailers are considered fairly safe in an accident.

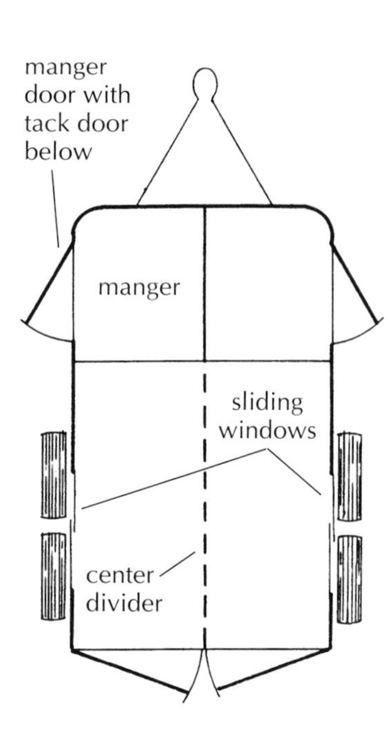

Straight-Pull, Two-Horse Straight-Load with two rear doors, two manger doors, two under-manger tack compartments, and two sliding windows

Straight-Pull, Two-Horse Straight-Load with walk-through to tack room, two rear doors, tack room door, two drop-down manger doors, and two sliding windows on sides and two on rear doors

Gooseneck, Two-Horse Straight-Load with tack room and two rear doors, two manger doors, two tack compartments, tack room door, and eight sliding windows

Slant-Load Trailers

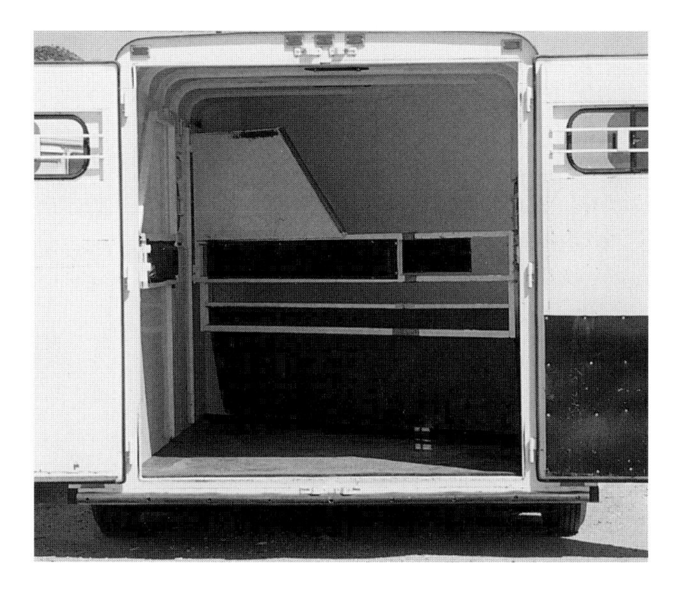

2.11 Slant-Load Trailer

Slant-load trailers are a relatively recent innovation. In a slant-load, the horses ride diagonally. It is thought that horses are more comfortable with acceleration and deceleration if they stand at an angle to the line of travel. If you were standing on a hay wagon, to keep your balance, you would likely not choose to face straight forward, but instead would stand somewhat sideways to absorb the jolt as the tractor or horses started and stopped. In a similar fashion, a horse's diagonal pairs of legs take turns resisting the various forces of acceleration and stopping. The slant-load design results in a shorter but slightly wider trailer. In the event of an accident in which the trailer is on its side, it can be difficult to get the horses out of a slant-load without righting the trailer or cutting off the top.

2.12 Collapsible Removable Saddle Rack

The odd triangle of space that is left over at the back of a slant-load trailer is often wasted space. You could opt for a collapsible, removable, or swing-out saddle rack in that area, as shown here. A permanent rear tack room (such as in photo 3.5) cuts the rear entrance in half. It eliminates the wide entry, which is one of the slant-load's most inviting features to a horse.

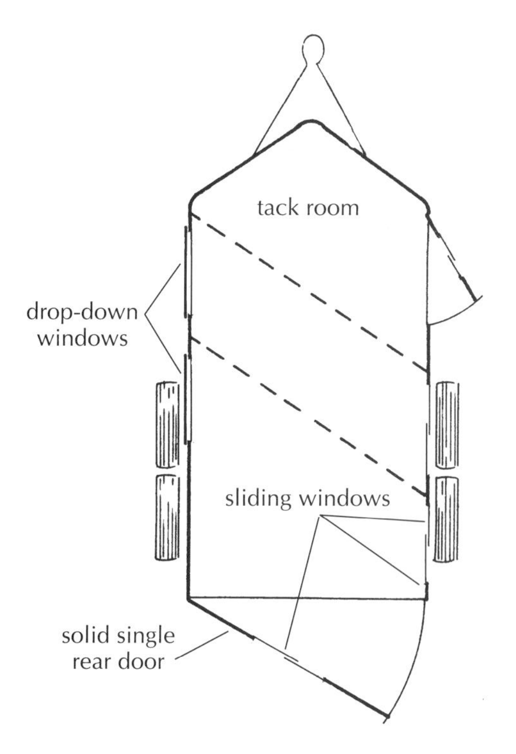

Straight-Pull, Two-Horse Slant-Load with tack room, one rear door, and a tack room door

Slant-Load Trailers (continued)

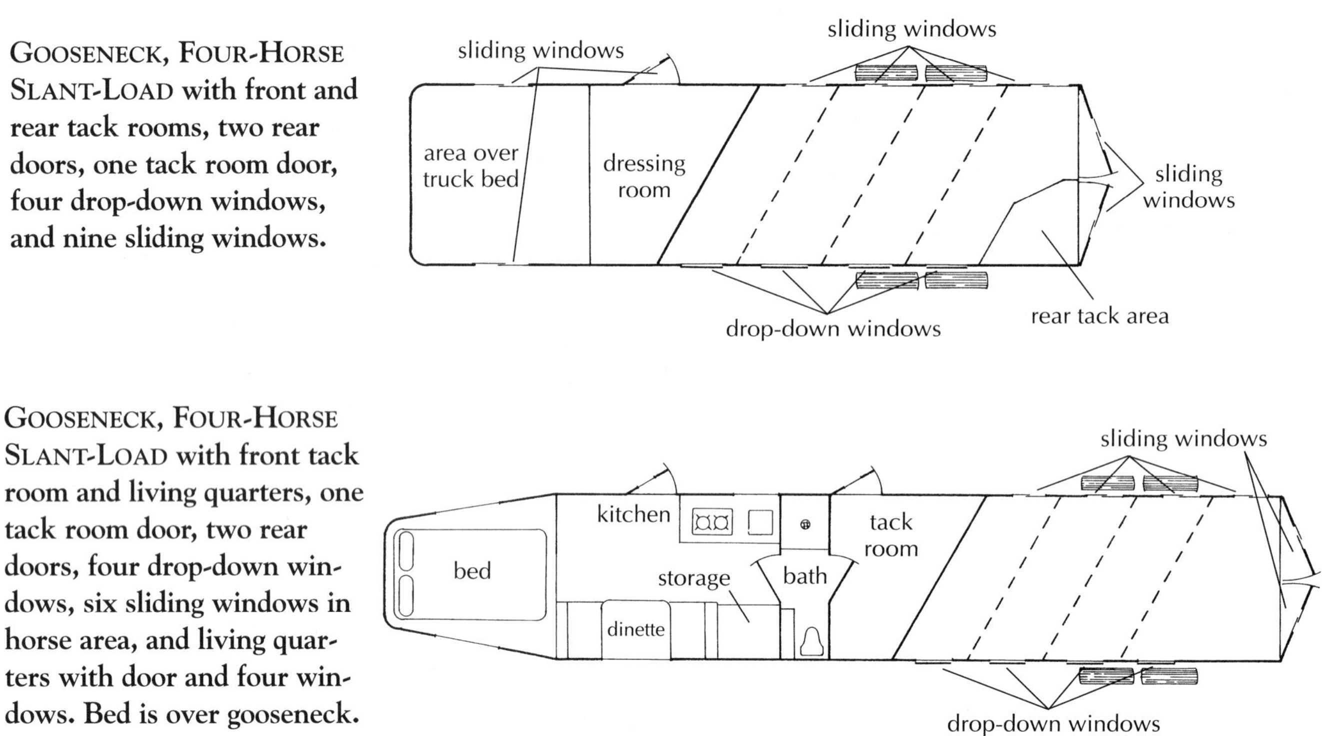

Gooseneck, Four-Horse Slant-Load **with front and rear tack rooms, two rear doors, one tack room door, four drop-down windows, and nine sliding windows.**

Gooseneck, Four-Horse Slant-Load **with front tack room and living quarters, one tack room door, two rear doors, four drop-down windows, six sliding windows in horse area, and living quarters with door and four windows. Bed is over gooseneck.**

Pros and Cons of Straight-Load, Stock, and Slant-Load Trailers

Trailer Type	Pros	Cons
Straight-load	Most common, largest selection in used market; lowest priced; often has a manger to feed hay and grain to horses; easy access to both horses; same width as towing vehicle so tires follow truck tracks.	Restriction of forelegs by manger wall; side-wall scramble to maintain balance; horse must do majority of bracing on front legs; hay always in horse's face; with manger horse can't lower head to clear respiratory tract.
Stock	Open, inviting; comparatively lighter in weight; less expensive; horse can find most comfortable way to stand; horse can be turned around and unloaded walking forward; versatile, useful for hauling hay and equipment.	Horse can get dusty or chilly from open sides (sides on some can be closed with Plexiglas inserts).
Slant-load	Shorter (1–4 feet [0.3–1.2 m], depending on trailer); open, inviting; some have option of turning horse around to unload; if tied with long lead, horse can lower head; useful for hauling hay and equipment.	Slightly (1–2 feet [0.3–0.6 m]) wider trailer, so it tracks differently from a truck and trailer tire might hit curb or drop off edge of road if not careful; more expensive than straight or stock; usually best suited for small to medium-size horses.

COURTESY TURNBOW TRAILERS, OILTON, OK

2.13 Reverse-Load Trailer

Reverse-load trailers have both a front and a rear entry. The horses load in one door and go out the other without having to turn around or back up. A horse can be loaded facing forward or rearward. Reverse-load trailers are specially engineered and balanced for horses to ride either way. *Never* load your horse facing rearward in a trailer unless the trailer is designed to be used as a reverse-load trailer. If you load your horse facing the rear in a conventional two-horse trailer, the horse's weight will put the trailer off-balance to the rear, which will result in trailer sway.

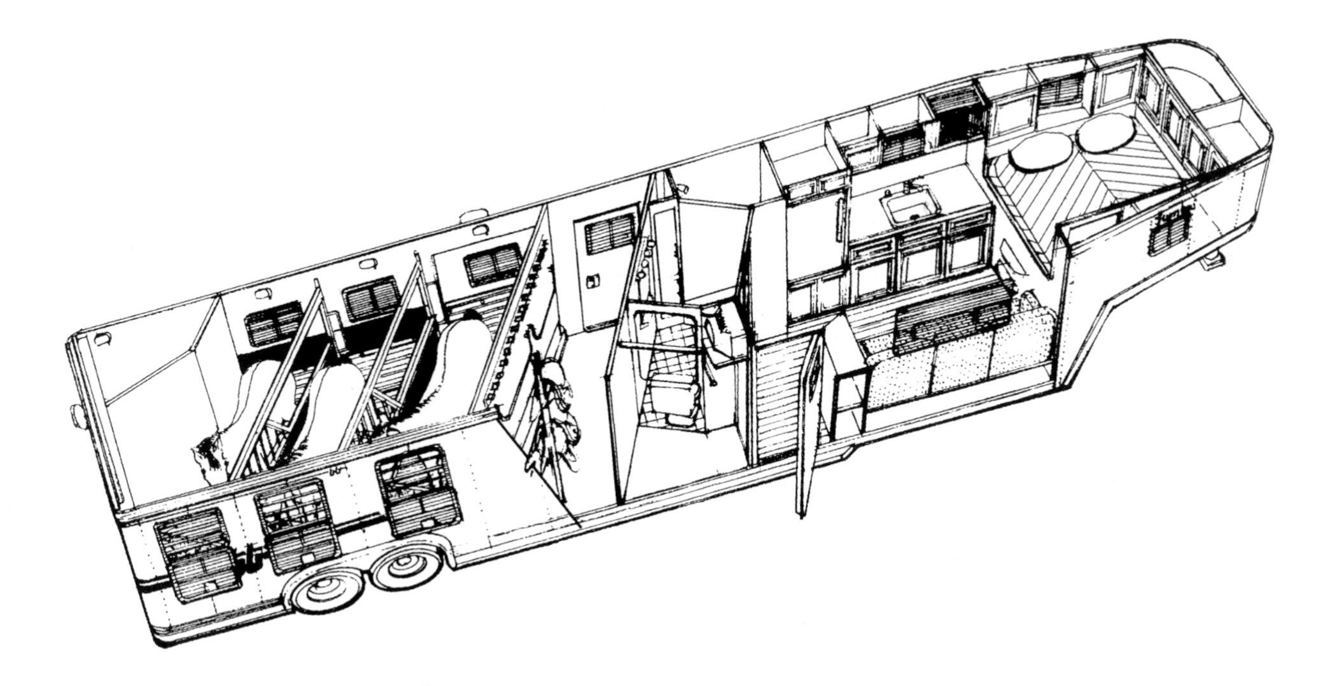

Reverse-Load Living-Quarters Trailer with three stalls, a tack room, and bath, kitchen, and bedroom. (Courtesy Turnbow Trailers, Oilton, OK.)

Loading Entry

Horses load by stepping up into the trailer or by walking up a ramp.

The step-up is the simplest, least expensive, and most common style of entry (photo 2.14). The open doors can provide wings to help guide a horse when loading. On high step-ups, a horse could bang a shin on the floor sill when loading. When unloading, he could catch a foot underneath. Take care when unloading horses in wet weather and on slippery ground or grass: That's when a horse's hind leg is most likely to slip under the trailer sill.

Some step-up trailers are difficult for a short horse to negotiate, especially if more than 16 inches (41 cm) off the ground. A ramp would make loading easier for a lame or old horse (photo 2.15). Ramps also allow you to conveniently load a garden tractor or an appliance on a dolly.

However, a ramp adds more expense and substantial weight to the rear of a trailer, where you really don't want it. A one-piece ramp adds an extra 200 to 300 pounds (91 to 136 kg). (See also photo 3.10). Ramps can spook an animal. Some horses react with distrust at the sight of a ramp, others startle when they hear the hollow sound or feel the movement of the ramp when they step on it. With a ramp, there are often no lower trailer doors to act as wings when loading. A horse that is evading loading could duck under the top door and scrape his back. Or he could simply step or fall off the ramp when loading. The area where the ramp hinges with the floor catches debris, which must be cleaned out before raising the ramp. Also, the crevice is a potential danger for a small hoof. The length of the ramp and the height of the trailer floor will dictate the slope the horse must climb. Ramps of 40 inches (102 cm) make for a steep climb. One-piece ramps can be very heavy and hard to lift. A power-assist or a very well-balanced spring device will make it easier for you to close the ramp (door). But you still must bend down and pick up the ramp. The ramp's hinge and the spring or power-assist add more moving parts that can require repair. Also, the plywood under the rubber mats on ramps must be inspected and replaced when necessary.

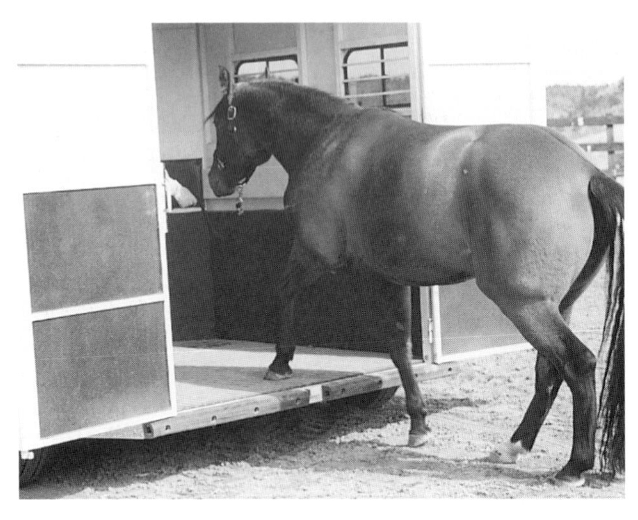

2.14 Step-Up

For safety, choose a trailer with a 16-inch (40.6-cm) step-up height or less. This one is 12½ inches (32 cm). All step-up trailers should have a rubber bumper (shown here) or a round steel bumper that runs the entire width of the rear of the trailer.

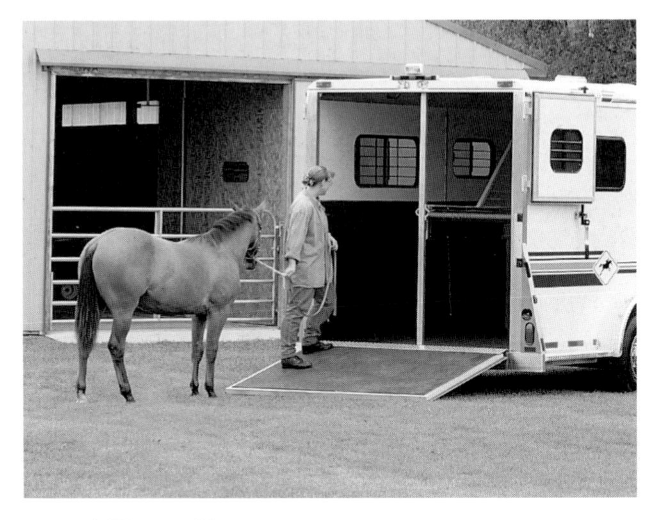

2.15 Full Ramp

A well-constructed, one-piece ramp makes it easy to load young horses into tall trailers. The gradual incline of a ramp 48 inches (122 cm) or longer makes it a simple matter for a small horse to load.

2.16 Individual Ramp
Individual ramps are lighter and easier to pick up but are also narrower, making it more likely for the horse to step or slip off. Note that when a trailer is not parked on level ground or when the ramp is warped, the ramp will not sit flat on the ground. The first hoof the horse places on this ramp will cause the ramp to depress and move, which frightens some horses.

2.17 Side Ramp
Side ramps located at the front of a larger trailer are handy for unloading the front horses: Just walk the horses forward. They're also good for loading horses in trailers where the animals ride facing backward. They are especially appropriate for four- and six-horse models, so you can lead some of the horses out the front, and unload the rear horses by backing them out of the rear of the trailer.

Trailer Materials

Materials commonly used in the construction of trailers include steel, fiberglass, and aluminum.

A trailer with a frame and skin of steel is sturdy but can be heavy, requires painting every five years, is subject to rust, and the lower interior walls will eventually need to be reinforced or replaced because of damage from horses' hooves and shoes (see photo 3.3).

Fiberglass is sometimes used for roofs and fenders, as it is cool, lightweight, and easy to repair.

Fiberglass-reinforced-plywood (FRP) trailers make up a small percentage of the market. They are built on a steel or an aluminum frame. The skin is fiberglass-covered plywood. The resulting trailer is the weight and price of steel but, if quality FRP is used, has lower maintenance requirements than does steel.

2.18 Steel Trailer
The short two-horse steel trailer with under-manger tack compartment was the most common horse trailer configuration for many years, and is still a popular, economical choice. This two-horse straight-load, step-up model has an escape door (the largest door on the side of the trailer; see page 37), a manger door (the small door on top near the front), and a door to the under-manger tack storage (lower door). With such a tack compartment, you have to bend over and reach inside. You'll have room for a saddle on each side, plus a few bridles and halters.

2.19 ALUMINUM TRAILER
All-aluminum trailers weigh the least, cost the most, and last the longest. Almost half the new-trailer market consists of aluminum trailers. An all-aluminum trailer, such as this one, has an aluminum frame and skin. It costs 30 to 50 percent more than an equivalent steel trailer, is 20 to 30 percent lighter, requires very little maintenance, and holds its resale value better than any other trailer. There is no need to repaint, but aluminum dents more easily than steel. Damage to an aluminum trailer will likely be greater than to a steel trailer in the event of an accident.

Steel-frame, aluminum-skin trailers weigh less than all-steel trailers, eliminate the need for painting, and decrease rust problems, but are almost double the price of a steel trailer. Because of a type of corrosion that occurs between steel and aluminum, trailers with a steel frame and an aluminum skin require special fastening techniques that keep the two metals from touching. (See photo 2.19.)

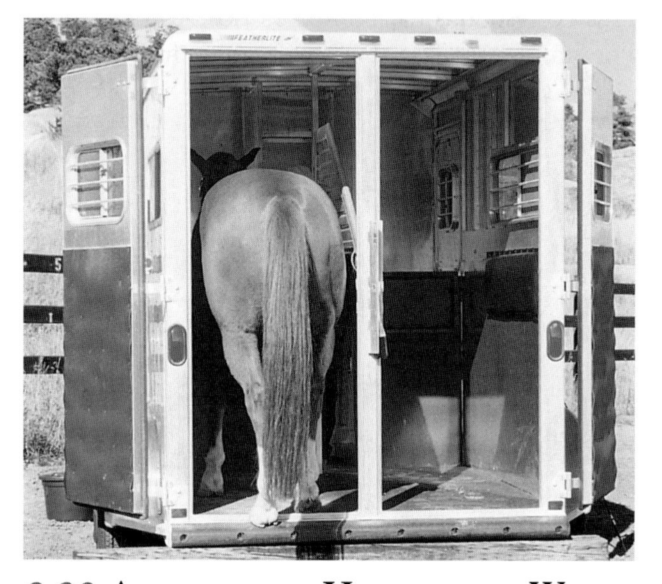

2.20 APPROPRIATE HEIGHT AND WIDTH
This 33-inch-wide (84-cm), 86-inch-tall (218-cm) trailer is of ample size for this 16-hand, 1,200-pound (544-kg) gelding.

Trailer Specs: Height, Width, Length, and Weight

When choosing a trailer, pay attention to the interior and exterior height, width, and length of the trailer, and choose a model that is appropriate for the type of horses you will be hauling. Be certain that the trailer's weight is something your towing vehicle can handle safely.

Height

The height inside an old-style horse trailer is about 78 inches (198 cm). New, standard-size trailers range in inside height from 80 to 86 inches (203 to 218 cm), but it is not uncommon to find Warmblood and draft horse trailers as tall as 96 inches (244 cm). Be aware that the height at the entry might be 4 inches (10 cm) lower than that at the ceiling, and the roof's support ribs are a height somewhere between the two. Whether you are shopping for a new or a used trailer, always take along a tape measure. Also, know the outside height of your trailer, including open vents and roof rack, so that when you pull under a marked overhang, such as at a motel, you don't rip off the roof of your trailer.

Width

The standard stall width is 26 to 38 inches (66 to 97 cm). The stalls in slant-load trailers are generally narrower than those in straight-loads. Warmblood trailer stalls can be as wide as 48 inches (122 cm). The average 15- to 16-hand saddle horse does well in a standard-width stall.

Length

The stall length of a straight-load is measured from the breast wall of the manger to the butt bar and corresponds to the horse's chest-to-tail measurement. The length of standing room in a straight-load ranges from 66 to 80 inches (168 to 203 cm). The average 15- to 16-hand horse measures about 66 to 68 inches (168 to 173 cm) from chest to tail, so a stall 72 to 78 inches (183 to 198 cm) long (from manger to butt bar) works well for most horses.

Measure the length of a slant-load stall down the middle of the stall from one wall to the other. This corresponds to the horse's nose-to-tail length. The stall length in a slant-load ranges from 90 to 96 inches (229 to 244 cm). A 15- to 16-hand horse is 92 to 94 inches (234 to 239 cm) from nose to tail, when the horse's head and neck are in a natural, relaxed position. Many slant-load stalls are too short for horses over 16 hands.

Weight

The standard curb weight of a trailer can be supplied by the trailer dealer. The actual curb weight includes the weight of any nonstandard add-ons such as 300 pounds (136 kg) for a rear ramp, 200 pounds (91 kg) for sidewall mats, and so on. To determine actual curb weight, weigh your trailer empty on a commercial scale (see chapter 1). Stop with just the trailer wheels on the scale, block the trailer's wheels, and detach the trailer coupler from the truck's receiver. If you don't detach the trailer, part of its weight (tongue weight) is being borne by the truck, and you won't get an accurate measurement.

Gross Vehicle Weight (GVW; also called Loaded-Trailer Weight [LTW]), is the weight of a trailer and all of its cargo, including horses. To determine GVW, weigh the loaded trailer on a scale. To estimate GVW, add curb weight, weight of horses, plus 100 pounds (45 kg) per horse for feed, water, and tack; 1,000 pounds (454 kg) is often used as a standard horse weight, but know your horse's actual weight. The GVW should never be greater than the trailer manufacturer's Gross Vehicle Weight Rating (GVWR) for the trailer, nor should it be greater than the truck's Maximum Trailer Weight Rating (MTWR).

Head Room

A 16.2-hand horse stands 66 inches (168 cm) at the withers. If you were to load him into a 78-inch (198 cm) trailer, he would be able to raise his head only 12 inches (31 cm) above his withers when loading or unloading, and would have to carry his head that low throughout the trip. A 16.2-hand horse would be much more comfortable in an 84-inch-tall (213-cm) trailer. (See chapter 8 for more photos showing horses inside trailers.)

Sample Curb Weights of Trailers

Trailer Type	Weight in lbs (kg)
Two-horse aluminum straight-pull, straight-load, with dressing room	2,500 (1,134)
Two-horse aluminum straight-pull, slant-load, with dressing room	2,800 (1,270)
Two-horse steel straight-pull, straight-load, tack compartments	3,000 (1,361)
Two-horse steel straight-pull, straight-load, with dressing room	3,300 (1,497)
Two-horse steel straight-pull, slant-load, with dressing room	3,800 (1,724)
Four-horse aluminum gooseneck, straight-load, with dressing room	4,500 (2,041)
Four-horse aluminum gooseneck, slant-load, with dressing room	4,750 (2,155)
Four-horse steel gooseneck, straight-load, with dressing room	7,000 (3,175)
Four-horse steel gooseneck, slant-load, with dressing room	7,500 (3,402)

Trailer Laws

State and federal trailering and trucking laws are designed for the safety of you, your horses, and other motorists. It is your responsibility to know the laws and adhere to them — fines can be stiff.

State Laws

Be sure you know the laws of the states in which you will travel, and pay close attention to:

- Maximum speed limit: 55–75 mph (90–120 km/hr)
- Maximum trailer length: 33–60" (84–152 cm)
- Maximum trailer width: 8'–8'6" (2.4–2.6 m)
- Maximum trailer height: 13–14' (4–4.3 m)
- Maximum length truck and trailer: 55–85" (140–216 cm), all of which may vary.

Every state requires separate trailer brakes for trailers more than 3,000 pounds (1,361 kg), which includes all horse trailers. Every state requires stop-lights, taillights, license light, and turn signals. Almost all require safety chains, breakaway brakes, clearance lights, reflectors, and insurance. Many states require you to carry flares. Check with your State Transportation Department or your motor club (such as AAA) for current regulations, or purchase a copy of *Motor Carrier Safety Regulations* at your favorite truck stop.

Factors Affecting Curb Weight

- **Style.** All other things being equal, a stock trailer is lighter than a slant-load, which is lighter than a straight-load.
- **Skin.** Aluminum is lighter than steel and FRP, which are about equal in weight.
- **Frame.** Aluminum is lighter than steel.
- **Roof.** Fiberglass is lighter than aluminum, which is lighter than steel.

Items That Add to the Curb Weight

- **Floor mats.** The type and number of mats will affect the trailer weight.
- **Sidewall mats.** Sidewall mats in a two-horse trailer can add up to 200 pounds (91 kg) to the curb weight.
- **Ramp.** A ramp adds 200 to 300 pounds (91 to 136 kg).
- **Spare tire.** A spare tire adds 40 to 60 pounds (18 to 27 kg).
- **Accessories in the tack room.** Carpeting, saddle racks, bridle racks, tack trunk, brush bins, and other accessories add considerable weight.

Federal Trucking Laws

The Federal Commercial Motor Vehicle Safety Act requires all states to follow the same minimum criteria for commercial drivers. States may have more stringent criteria than those of the federal laws, and each state and law enforcement official will interpret the law in a slightly different way. But basically, if your rig's GVWR or Gross Combined Weight Rating (GCWR) is 10,000 pounds (4,536 kg) or less, you do not have to be concerned with this law. If your rig's GVWR or GCWR exceeds 10,000 pounds (4,536 kg) but is under 26,000 pounds (11,793 kg), and you are using it for recreational purposes, the federal act does not apply to you. However, if it is over 10,000 pounds (4,536 kg) and you are using the rig for commercial purposes, you will be required to follow federal commercial vehicle requirements, which include obtaining a commercial driver's license and keeping a logbook. Commercial use includes hauling for hire, buying and selling horses, breeding, and doing things with horses that contribute to income. If your rig weighs more than 26,000 pounds (11,793 kg), no matter whether you are a recreational or a commercial user, you must follow federal commercial vehicle requirements.

3 ✦ TRAILER FEATURES AND OPTIONS ✦

If you are at the point where you know what you want — for example, an aluminum step-up, straight-pull, straight-load or a steel gooseneck slant-load with a ramp — you've settled some of the large issues but your decision making is far from over. There are many trailer details yet to consider. Some features are standard, others cost exra, and some options are not available on every trailer.

Trailer Construction

The workmanship and materials dictate, to a large degree, the cost of a trailer. Quality workmanship will be evident in the straightness of the frame, the fitting of seams, the finishing of edges, and the paint job.

A top-notch paint job not only improves a trailer's appearance but protects it as well. Progressive trailer manufacturers have taken advantage of the recent technological advancements made in paint quality and bonding. White is the most popular color for both the inside and the outside of a trailer, and for good reason. White absorbs the least amount of light and heat, and a white interior is inviting to a horse. Dark colors not only are heat absorbing but they also look like a dismal cave to a horse's eye. So before you order a custom red or black trailer to match your new truck, consider a white trailer for the comfort of your horse. A white roof is especially important.

All interior and exterior surfaces should be flush whenever possible, and all edges should be finished round or covered with rubber or vinyl edging. Bolts and fasteners should have rounded, smooth heads or be flush with the surface they are on. Smooth interiors from top to bottom will prevent additional injuries in the event of a trailer rolling in an accident. Options that add to a trailer's life and appearance include gravel guards and undercoating. Other handy options are a spare tire mount, wheel covers, hay rack and ladder, and water tanks.

Don't assume just because a trailer looks good that it is safe. If you are new to trailer buying, get competent advice from an experienced horseman. Unless you're very knowledgeable, instead of going bargain hunting, you're better off buying a brand-name, well-respected trailer from a reputable dealer that also offers service. Top-of-the-line trailers will often carry a 5-year warranty. Study the warranty carefully and then bone up on trailer and towing-vehicle maintenance.

Brakes

A separate trailer brake system with a manual lever within reach of the driver should be an integral part of any horse trailer rig. In all states, they are required for trailers over 3,000 pounds (1,361 kg), which includes all horse trailers. In most instances, brakes are required on each axle. A four-wheel brake system is safer and more efficient than is a two-wheel brake system, even for a two-horse trailer, and four brakes are imperative for larger rigs. State laws dictate which trailers require brakes and whether brakes are required on all four wheels. (See the appendix for information on state trailer brake laws.)

Although electric brakes are most common, some manufacturers offer the option of vacuum-hydraulic brakes.

Electric trailer brakes are wired through your truck's electrical system and are activated when you step on the brake pedal, use the hand controller, or when the emergency trailer breakaway system is activated.

Hydraulic brakes are independent of the towing vehicle and are activated when pressure is exerted on the coupler of the hitch, such as when you are slowing down. The system is controlled by a master hydraulic cylinder located at the front of the trailer.

Emergency Breakaway Trailer Brakes

Federal traffic safety law requires that all trailers that require brakes have an emergency breakaway braking system.

The breakaway brake mechanism is battery-operated and designed to activate your trailer brakes if your trailer should happen to come unhitched from the truck. The system's battery is usually located in the tack room or on the tongue of the trailer. The switch that activates the brakes is generally on the tongue or neck of the trailer. Plugged into the switch is a plastic breakaway pin with a cable attached to it. When you are towing, the other end of the cable should be securely attached to the frame of your truck. If the trailer were to separate from the truck, the cable would pull taut, and the pin would pull out of the switch and activate the trailer brakes. In order to work properly, the breakaway cable should be slightly longer than the safety chains. That way, if the trailer hitch breaks or comes uncoupled but the chains still hold, you can use the manual trailer brake controller to gradually apply the trailer brakes, rather than having the brakes lock up suddenly and possibly snap the safety chains.

Never leave the pin out of the switch any longer than necessary, because while it is out electricity is powering the brake magnets. In a matter of minutes, the battery will be completely drained. For a step-by-step sequence for hooking up the emergency breakaway cable, testing, and maintenance, see pages 49 and 116.

3.1 **Here is an emergency electric trailer brake switch on a gooseneck with the cable plugged in. Note the battery strapped in place on a shelf above the spare tire. Both items should be locked in place to prevent theft. (For more goosenecks, see photos 2.7 and 3.24.)**

Roof

It is best if the trailer roof is one piece. Two- and three-piece roofs invite leakage and often require recaulking.

Suspension

Good-quality trailer suspension is sturdy but not stiff. Horse trailers usually have either leaf springs (old-style trailers) or rubber torsion bars (newer models). Rubber-torsion suspension is said to give a better ride than leaf-spring suspension. Rubber-torsion suspension incorporates large sections of dense rubber to minimize bumps, vibration, and noise, resulting in a quieter, smoother ride on highways. Leaf springs deliver a stiffer and somewhat noisier and rougher ride, but they will absorb the drop from a deep hole or a rut better than will rubber torsion, so they are good for ranch roads, travel in the back country, and across pastures. A suspension check should be a part of the routine maintenance of your trailer (see chapter 4).

Flooring

Most steel trailers and some aluminum trailers have wood floors. Oak or pine is fine as long as it is No. 1 quality wood. Usually 2 x 6 or 2 x 8 boards are used, and they should be of a consistent thickness, straight grained, and have no knots, twists, or warps. Many aluminum trailers have aluminum floors with holes at the rear for drainage of urine.

3.2 Pressure-treated wood will resist rot from manure and urine longer than will nontreated wood. The pressure-treated pine floor in this slant-load trailer has been painted white for additional protection and ease of cleaning.

Mats

Cover the floorboards with rubber mats. Removable rubber mats at least ¾ inch (1.9 cm) thick will provide cushioning, traction, and protection for floorboards. Remove the mats regularly to let the floor dry. Half-inch (1.3-cm) mats, which are standard in some trailers, are easier to remove but don't last as long or provide as much cushion for the horse as ¾-inch (1.9-cm) mats. One-inch (2.5-cm) mats offer more cushioning but are very heavy.

Sidewalls

Sidewalls come in single, double, and triple layers. Double and triple walls add more weight to the trailer, but they provide insulation that makes a trailer cooler in summer and warmer in winter. Insulated walls are sturdier and they decrease road noise. Any sidewall should be thick enough and of a material that will resist the destructive forces of a horse's hooves.

3.3 The sidewalls of this old steel trailer show much scarring from scrambling hooves. The steel lining has rusted and is now buckling away from its attachment at the floor, making a dangerously sharp edge for a horse's lower legs. Most steel trailers eventually need sidewall reinforcement or replacement. (See photos 3.11, 3.12, and 4.1.)

Tack Room/Dressing Room

A tack room can range from a small compartment under the manger for just a saddle and bridle to an extensive walk-in dressing room. A walk-in tack room/dressing room adds about 4 feet (1 m) in length to a trailer. It should have doors that lock securely. A door on the driver's side is most convenient for quickly grabbing a halter, and when you saddle your horse, you'll be carrying the saddle directly from the tack room to your horse's near side. All items in the tack room should be securely stowed and all doors should be locked: In the event of an accident or sudden stop, the doors won't come open, allowing items to fly into the horse compartment or onto the roadway.

COURTESY 4-STAR TRAILERS, OKLAHOMA CITY, OK.

3.4 Tack Compartment
This two-horse, straight-load aluminum trailer has a tack room in front (door closed); a tack compartment under the manger, complete with a saddle rack and its own lockable door; a manger door; and a half-escape door.

3.5 Rear Tack Room
This slant-load trailer has a permanent rear tack room. Notice that the center post has cut the entrance in half, making this a narrow loading opening, as on a straight-load. This means that you either have to enter ahead of your horse or must send the horse in first and then follow him in to tie him. Not good.

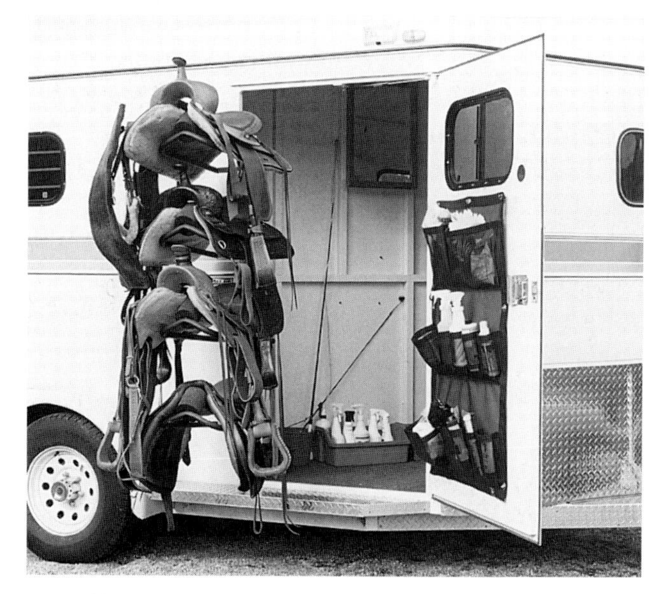

3.6 Dressing Room
This tack room/dressing room is a walk-in room at the front of the trailer. It features a swing-out saddle rack and handy door organizer.

Doors

Doors and latches should work easily. Slam latches are the handiest, and flush handles are the safest. Be sure all doors and latches are strong and that they fit squarely, or they might be difficult to close or latch.

Escape doors come in many styles and sizes, but they are not designed for "escaping." If you feel you might have to escape when you're loading your horse, you're taking a big risk! Escape doors should really be called "access doors," as they allow you access to your horse so you can check on him from the side or the front. It is true that if you have an escape door on your trailer, you can lead the horse into a stall by walking in ahead of him and leaving via the escape door. But there are two risks involved in this method. Anytime you walk into a trailer ahead of your horse, you risk getting trampled if the horse suddenly lunges in. And, of course, you don't have the training control like you do when you are working alongside him. Also, when using an escape door, some horses try to follow the handler right out the door and may panic or become wedged and suffer serious injury.

If you have a walk-through door to a front tack room, you can lead the horse into the trailer by going into the stall ahead of him and leaving via the tack room. Again, the risk of being trampled if the horse rushes in is a real possibility. For an example of a walk-through-to-tack-room layout, see illustration on page 24 and photo 3.8.

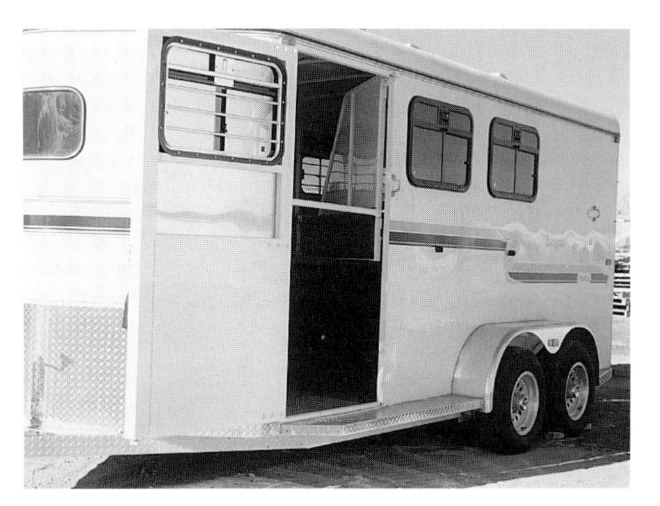

3.7 Full-Size Escape Door

The escape door on this slant-load has a hinged bar across its middle to discourage the horse from coming out. You might be tempted to unload your horse through this door, but that would be dangerous because of the diamond plate running board, which is slippery and has a hard, sharp edge. A door like this can be used to unload a horse in an emergency, however.

Rear Doors

Rear-door configuration is usually either two double doors (top and bottom), two single doors, or one wide door. Although leaving off the top half of a double door or keeping it open when traveling can provide extra ventilation on extremely hot days, it is generally not a good idea. Your horse will probably be filthy and stressed from all the debris and traffic noise that are sucked into the trailer like a vacuum. Cigarette butts from your vehicle or others could be drawn into the trailer and cause a fire in the bedding or hay. Rear doors with windows and screens are a better option.

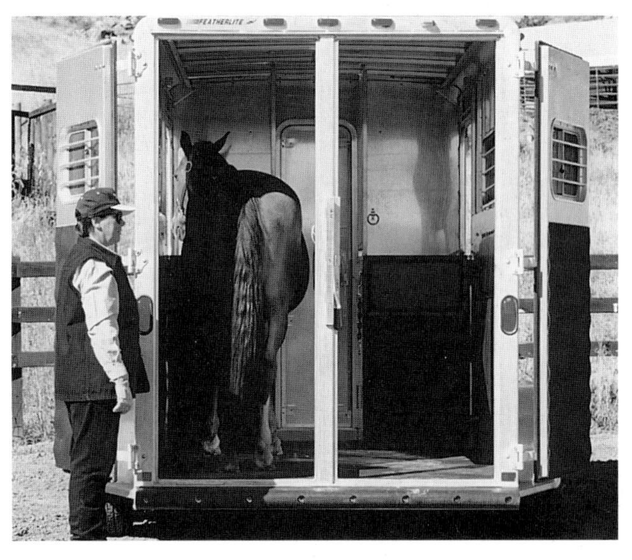

3.8 Double Rear Doors with Center Post

These are the most common doors for a straight-load trailer. The center post gives the stall divider a place to attach and adds strength and stability to the trailer. It can cause some horses to be reluctant to load if they feel confined by the post or bang into it when trying to load or unload. The post makes the trailer less versatile than a fully open rear entrance.

3.9 Double Rear Doors without a Center Post

These doors are common on slant-load trailers. The wide opening of the entrance is inviting to horses and gives you plenty of room to work, turn the horse around, and unload facing forward. In addition, the open doorway makes it easy to haul furniture and other large, bulky items.

When a slant-load has a permanent tack room at the rear, the resulting space makes a narrow loading entrance. (See photos 3.5 and 8.1.)

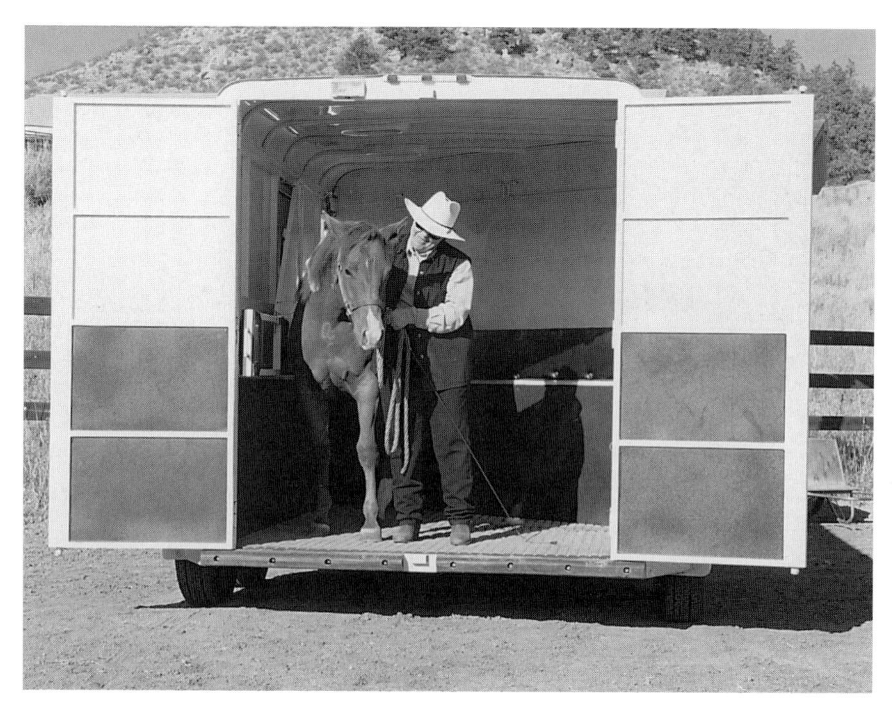

3.10 Double Rear Doors with Ramp

With the height of this gooseneck trailer, a ramp will make loading easier for some horses.

Latches

Poor latches have been responsible for horses falling out the back door of a moving trailer. Latches that are flush with the outside surface of the trailer are safer in terms of snagging horse or human.

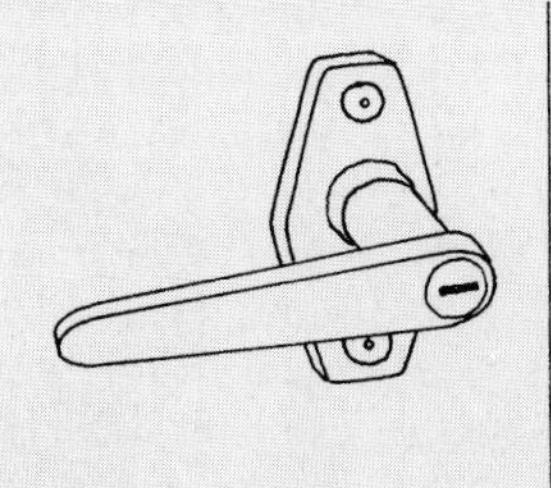

Handle Latch

A handle latch like this is a snagger anywhere on a trailer. It can result in injury to the horse and damage to tack, clothing, or the trailer itself. Flush, slam latches are much safer. (See photo 10.27.)

Rear-Door Bar Latch

The rubber-covered handles on this typical rear-door bar latch operate vertical steel rods located within the square tubing. The ends of the rods lock into pockets at the top and bottom of the trailer door frame. (See photo 3.22 for a close-up of this feature.)

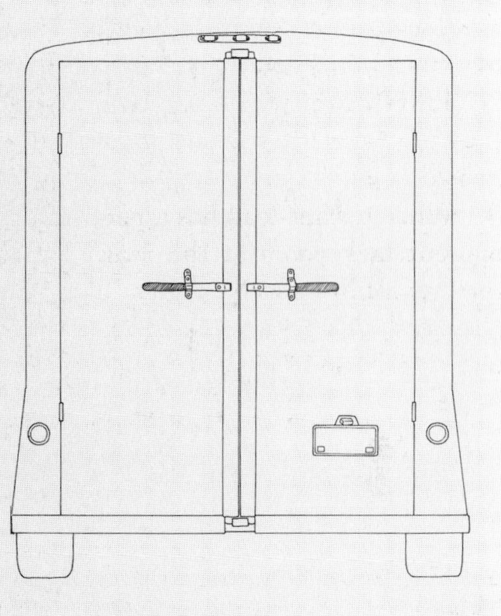

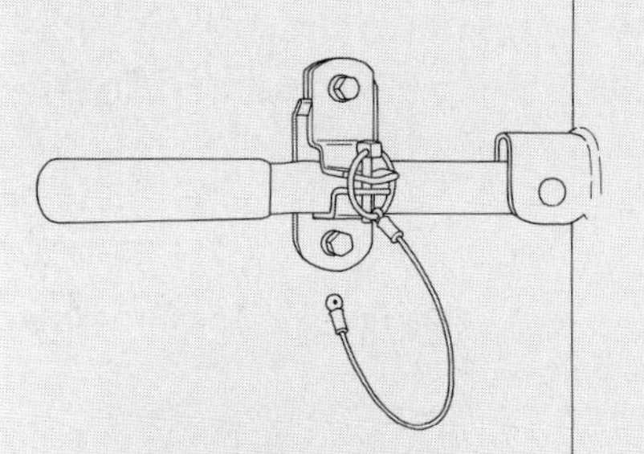

Pin-Type Ring Lock

For added safety, a pin-type ring lock can be clipped through the handle keeper of a bar latch to prevent it from coming undone in transit. When not in use, the ring lock is suspended by a metal cable riveted to the trailer door so it is always there when you need it.

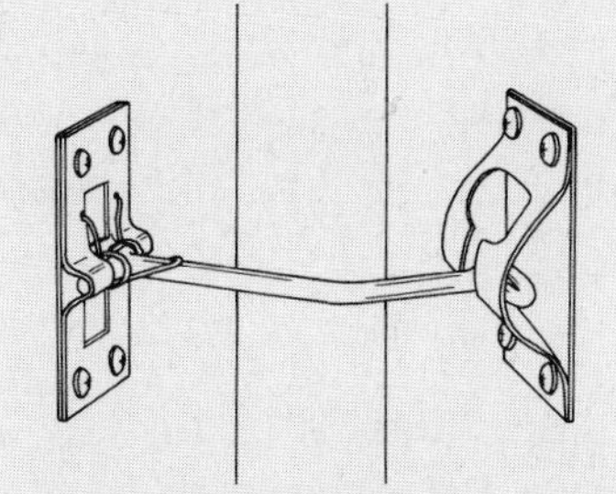

Hold-Back Fasteners

All doors should have hold-back fasteners. This latch on the back of a tack room door keeps the wind from banging the door shut just as you are walking out with an armload of tack. If your trailer doesn't have hold-back fasteners, you can innovate by using a bungee cord with hooks on the ends.

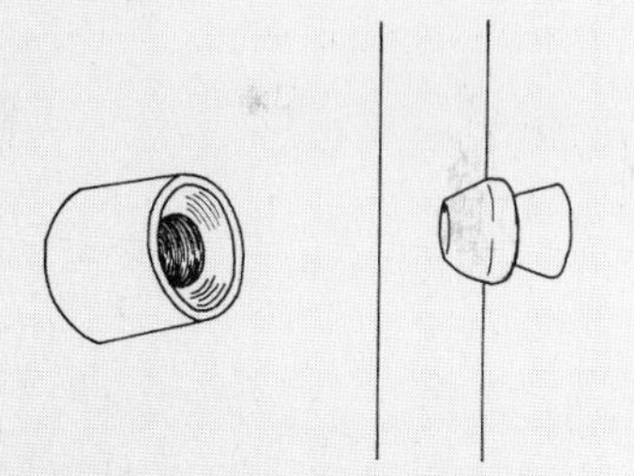

Rubber Dimple-and-Nipple-Type Fastener

The rubber dimple-and-nipple-type door fastener is popular on newer trailers and safer than the hard metal nipple and plugs that they are replacing, but they are not as secure in high winds as hold-back fasteners.

Butt Bar, Chest Bar

A butt bar should be located so a horse contacts it before he contacts the pressure on his lead rope. Many horses lean on a butt bar to help balance, so the device (and its fasteners) must be strong and located low enough for the horse to sit on it. Wrap the horse's tail to prevent damage from rubbing.

The fasteners of the bars at the chest and tail and at the center divider should all work easily. Some can be difficult to operate if the trailer is on less than completely level ground. Fastening and releasing the bars should be quick and easy — it is not something you want to have to play with or bang on. Be sure the bars can be unfastened when there is pressure on them.

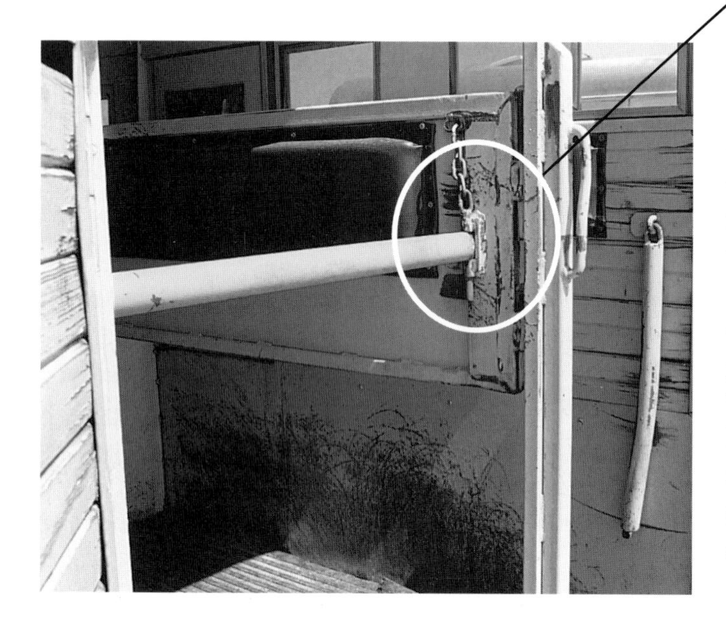

3.11 Metal Butt Bar

This typical straight-load metal butt bar hinges at the wall of the trailer and fastens to the center divider by dropping a keeper pin (which is suspended from a short piece of chain) through a three-part channel.

Getting the three parts to line up can be a challenge, especially if the trailer is not on level ground. The rust on this trailer makes questionable its strength and safety. Make sure the butt bars all are fastened whenever the trailer is moving. This will keep the bars from swinging and damaging the trailer walls, and will keep down the noise level for a horse traveling alone.

3.12 Padded Butt Bar

This metal pipe butt bar has been padded with foam and a vinyl covering. Notice that the keeper pin is attached to the trailer wall by a short cable rather than a chain. Note, too, that the lower wall of this straight-load has been reinforced with rubber matting.

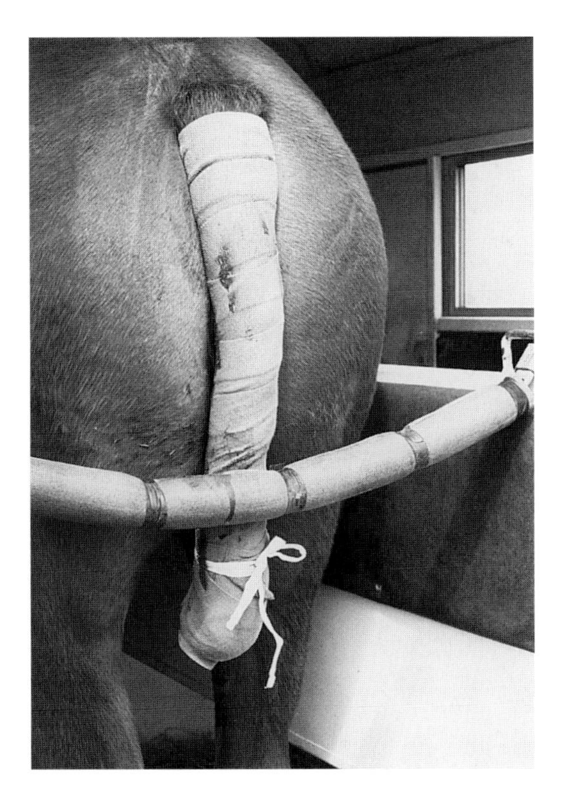

3.13 Padded Butt Chain

If the trailer has a butt chain, you can protect your horse's tail by covering the chain with a piece of foam pipe insulation. Here, the insulation is held in place with electrician's tape. This horse's tail has been wrapped completely to keep it clean in transit.

3.14 Chest Bar

This straight-load trailer has a chest bar instead of a manger. The design allows you to place the horse's feed ahead of him at a height that is comfortable for eating. Most importantly, this design allows a horse to lower his head and neck to blow and clear his respiratory tract. The horse's front legs do not bang into a manger wall, nor is there much chance of him getting tangled in a feeder or hay net. Each stall has an access door. You can move the center divider to one side or the other when loading or you can fasten it as shown here to give a single horse more room when traveling.

Stall Dividers and Padding

Some stall dividers are removable for hauling large or difficult travelers or to accommodate a mare and foal. Slant-load dividers usually fold flat against the wall. Most horses travel more comfortably when a divider goes no more than halfway to the floor. A rubber mat can be attached on the center divider, which extends almost to the trailer floor, to prevent two horses from stepping on one another. The head portion of the divider can be solid or bars. With a solid divider, there will be less playing or fighting; with bar dividers, there will be better airflow for comfort.

Padding on the stall fronts and sides, the center divider, and the back door increases safety and comfort. (See photos 3.11, 3.12, and 8.40 for examples of padding.)

Tie Rings

There should be at least one tie ring inside and one outside for each horse. The design of the tie ring should be strong and safe.

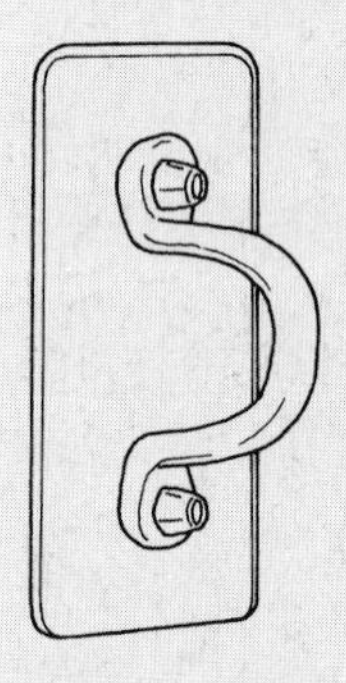

Outside Tie Ring
Outside tie rings should be located away from door latches, light fixtures, and anything else your horse could chew on or become snagged on. Tie rings should be welded to the trailer framework or secured by bolts that go all the way through the trailer wall. (See also photo 3.19.)

Inside Tie Ring
Inside tie rings, such as this straight-load manger ring, should be strong and mounted where they can be accessed easily from both inside and outside the trailer.

Windows and Vents

Horse trailers can get very hot. Windows and vents should be well designed and well placed. All glass or Plexiglas should be tinted to reduce glare and heat from the sun. Plexiglas is safer in the event of an accident but scratches more easily than glass. A minimum of one bus-style (sliding) window at the head and one at the tail (or side) of each horse is suggested. A third window per horse located on the back door is even better. It is best if you are able to open all windows from the outside but can lock them from the inside if desired. In addition, each stall should have at least one overhead vent.

3.15 Small Manger Window
This small manger window can only be opened when the vehicle is stopped. The only air flowing in from the front is through the small butterfly vent at the side of the horse's neck. This old-style vent can be operated from outside the trailer, which is handy, but its low location doesn't do much to remove warm, moist air from the trailer. In cold weather, it is located at a height that will deliver a chilling wind onto the horse. This is not a good choice.

3.16 Drop-Down Windows

The drop-down windows on this four-horse slant-load have sliding bus windows in them that can be left open for traveling. Although it is fine to open drop-down windows when the vehicle is stopped, *never* leave them open when traveling unless the openings are also outfitted with strong interior grilles that will keep the horses' heads inside the trailer.

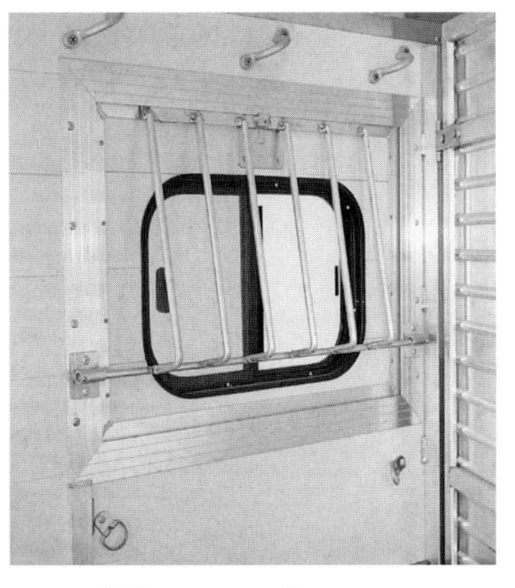

3.17 Window Grille

This heavy interior grille stays in place when the drop-down windows are opened, so you can fasten the windows in the open position for safe traveling in very hot weather. The grille has a separate hinge so it can be opened when the vehicle is stopped.

3.18 Roof Vent

Two-way vents are best. When you want air to flow in, such as during hot weather, the vent directs air in through the opening from the front. If you want to draw warm, moist air out of the trailer during cooler weather and you don't want a breeze to come in, the vent can be opened the other way to draw air out of the trailer toward the rear. Note the caulking around the vent and fasteners.

You want a minimum of one roof vent per stall. The interior trim on the vents should be smooth and rounded for safety. Because the vents open and close from the inside, if you are under 5 feet 6 inches (168 cm) tall, you might need a step stool to reach the vents in very tall trailers.

3.19 Sliding Bus Window

Sliding bus windows such as these should be located at the rear of each horse in a slant-load, alongside each horse in a straight-load, and in the rear doors whenever possible. Note the safe, sturdy tie ring.

Lights

There should be highly visible operating lights and clearance lights as well as interior lights in at least the manger and tack areas. Interior stall lights are useful when loading/unloading and when feeding at night. For night traveling, reduce the dramatic difference between the darkness and bright headlights for the horses by turning on the interior manger and stall lights. Especially with older trailers, you'll need to know which lights have their own separate battery source, which operate when the trailer wiring is plugged into the truck, and which require that the truck's ignition be turned on.

3.20 Interior Stall Light
This is an interior stall light at the rear of the horse in a slant-load. Note the permanent interior grille, which protects the bus-style window at the rear of the horse.

3.21 Weatherproof Light Switches
These weatherproof light switches are located next to the rear doors of this ramp-load trailer and are horse safe. Their low profile minimizes the chance of a horse getting injured on them. Note the reflector and taillight.

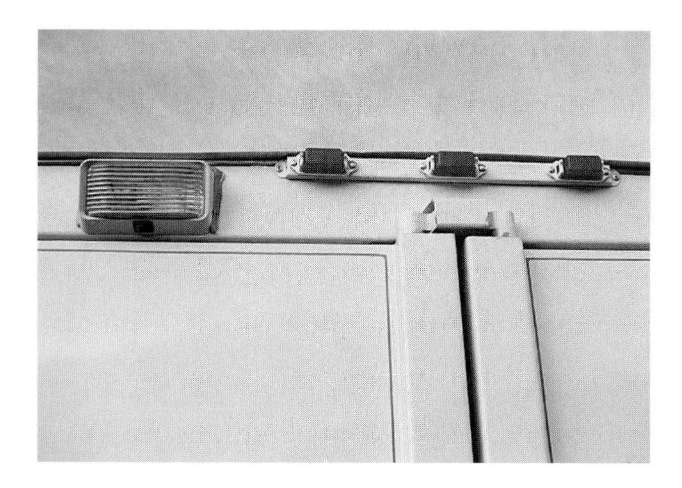

3.22 Rear Loading Light and Clearance Lights
This is an example of a rear loading light with a rocker switch underneath. The row of three red clearance lights is located above the midpoint of the trailer's rear door.

Tires and Wheels

Make sure your tires and wheels are the proper size for the trailer. For a two-horse trailer, they should be at least Load Range C (6-ply rating); for a three- to four-horse, they should be Load Range D (8-ply rating); for four-horse and over, they should be Load Range E (8- to 10-ply rating). Be sure that tires are inflated to their prescribed pressure, and check them often. (See chapter 1 for more about tires.)

Spare Tire

A spare tire may or may not be included when you buy a new trailer. Regardless, be sure you have one, and see that it is mounted on a wheel of the proper size and style for your trailer. Secure the tire where it can be locked to prevent theft.

3.23 Good Tires
These trailer wheels, hubs, and tires are in excellent condition. The tires are properly inflated and have plenty of tread. The hubs are smooth and rounded and would not injure a horse's legs if he were to bump into them.

3.24 Spare Tire
This living-quarters horse trailer shows spare tire, propane tanks for kitchen, and battery for emergency breakaway brake system.

Feeders

Many straight-load trailers have mangers that provide a place to feed hay or grain. (See photos 10.27 and 11.1.) The disadvantage with manger feeders is that the horse cannot lower his head and neck to blow, cough, and release dust and debris from his respiratory tract. Feed small amounts of very clean hay in transit. If using a manger-style trailer, let your horse get out every four hours to lower his head and blow. In straight-load trailers with chest bars and in slant-load and stock trailers, you can use hay nets, hay bags, or other types of feeders. (See photos 3.14, 3.26, and 10.26.)

3.25 Hanging Feeder
This hay bag (Fold-A-Feeder, Big Piney, WY) works in a slant-load trailer or in a straight-load with chest bar. It has a mesh back and bottom to allow hay fines and dust to drop to the floor. It is adjustable to various heights and can be folded and snapped closed when not in use.

Options

There are many options that will customize your trailer. A drip cap over the rear doors will send rain to the sides rather than down on you and your horse when you load or unload. Rustproof metal plates on the front and fenders of the trailer keep gravel and rocks from chipping the paint. (See photos 2.5, 2.18, 3.7, and 4.4.)

3.26 Roof Rack
The roof rack on this stock trailer makes a handy place to haul hay (as long as it's not raining!) and other large items. Fasten all items securely to prevent something from blowing or bouncing off onto the roadway or someone else's vehicle.

Buying a Horse Trailer

Now that you are familiar with many of the design choices and details available in horse trailers, put together a checklist of the important features and desired options for your situation. Whether you are buying a new or a used trailer, you'll need to know your answers to the following:

- Towing capacity of your towing vehicle?
- Straight-pull or gooseneck?
- Steel or aluminum?
- Straight-load, slant-load, or stock trailer?
- Number of horses?
- Size of horses being hauled?
- Tack compartment or dressing room?
- Living quarters?
- Your budget?

When buying a trailer, take your tow vehicle along so you can hook it up to evaluate the hitch or bed height, assess levelness, and check the lights and brakes.

Buying a Used Trailer

Used trailers generally have no warranty, so you or your mechanic should inspect the trailer carefully.

1. Bring a tape measure and make accurate measurements of inside and outside height, width, and length. Remember, most older-model horse trailers are narrower and shorter (78 inches [198 cm] inside height or less) because they were made for horses 14.2 hands and under. For a 16-hand horse, look for a stall that is 34 inches (86 cm) wide and 80 to 84 inches (203 to 213 cm) tall.

2. Remove the mats and check the floorboards for rotting, splitting, and warping.

3. Check the axles and frame for excessive rust, bending, or warping.

4. Don't just kick the tires; closely inspect the condition of the tires for weather checking (tire rot), which could necessitate replacement. The sidewalls of a stored trailer deteriorate from sun and moisture before the tread wears out, so check them carefully. Also look for uneven wear, which may indicate frame damage or faulty wheel bearings.

5. Ask to see the trailer maintenance log or receipts. If there is no record that the wheel bearings have been serviced, remove one or all wheels and brake drums and take a good look. Also inspect the brake pads and drums while the wheels are off.

6. Check for broken springs and worn bushings where the springs attach to the frame.

7. Check the hitch, coupler, and chains for cracks. Make sure everything is in good working order.

8. Examine the interior sidewalls, fenders, and front end for damage and rust. Fresh paint on a trailer may be hiding rust. Look for paint bubbles or a bumpy texture, which could indicate rust under the new paint. The best trailer color is white, inside and out, so if you are looking at a trailer that has been freshly painted black, find out why (better to cover rust?). Dark-colored trailers are very hot and uninviting during loading, so realize that if you buy a dark trailer, you will probably end up painting it.

9. Note whether the trailer needs a coat of paint and whether the trailer would need sandblasting to remove rust or body work to repair dents before painting.

10. Check all hinges and latches of doors and ramps to be sure they don't need repair or replacement. Again, be ever vigilant for rust damage.

11. Open and close all windows and vents and note any signs of damage from leakage around them.

12. Inspect the roof from the inside and from the top for signs of damage or leaking.

13. Hook up the trailer to your vehicle and test the running lights, turn signals, and brake lights, and make sure the brakes and emergency trailer brakes work (see chapter 4 for more about brakes). In all states, brakes are required on all horse trailers. Federal law requires an emergency breakaway brake system.

14. Ask the seller to accompany you while you take the trailer for a test pull.

15. Ascertain that the seller has a clear title to the trailer.

Remember, everything that doesn't work will have to be fixed for safety and to comply with regulations. Sometimes uncovering the cause of a problem, such as faulty wiring, takes quite a lot of time and money.

4

✦ TRAILER MAINTENANCE ✦

A trailer represents a big investment. If you take care of it, it should easily last 20 to 25 years. Each time you use your trailer, perform a pre-trip safety check as outlined in chapter 10 and an after-use follow-up as described in chapter 14. Fix or service anything that warrants attention at those times.

In addition to those procedures, design a routine maintenance plan for your trailer, just as you do for your towing vehicle (see chapter 1).

Examine structural components, such as floors, doors, and latches, on a regular basis to prevent catastrophes while on the road. Routine servicing of suspension, tires, wheel bearings, brakes, and other mechanical components not only prolongs the life of the vehicle, but also allows you or your mechanic to uncover problems before they become emergencies. Just because a truck and trailer operate fine for short local trips, that does not mean that the rig is roadworthy for longer hauls. Some of the most common problems encountered on long trips are overheated engines and transmissions, flat tires (truck or trailer), and problems with brakes, hitches, and lights. Service your truck and trailer regularly.

Check the following major items at least once a year and repair or replace as necessary.

- Registration
- Wheel bearings
- Brakes
- Lights and wiring
- Floor
- Breakaway brake
- Tires
- Doors
- Frame
- Body, waxing, or acid wash
- Suspension
- Latches, fasteners
- Trim

Registration

Annually, be sure to register and get new license plates or stickers on time, so your trailer is ready to use when you need it.

Wheel Bearings

The grease that lubricates wheel bearings picks up dirt and dries out. That's why you must have the wheel bearings cleaned and then repacked with grease annually or every 3,000 miles (4,839 km). Make sure that the seals, which keep out dirt and moisture, are replaced at the same time.

Brakes

Brakes should be inspected on a new trailer after 200 miles (323 km) of travel, and then serviced or adjusted every 3,000 miles (4,839 km). For a brake check, the wheels and brake drums should be removed. If you use an air compressor to blow accumulated dust and dirt out of the brake mechanism, be very careful to avoid breathing the dust, as old brakes may contain asbestos. You can instead wipe the inside of the brake drums and the surface of the brake pads with a damp cloth. Have the pads checked for wear, and replace them if necessary. If the brake pads have worn too thin, the rivet heads may have scored the drums — that's often the squeak you hear when the brakes are applied. (The squeak could also be from dust accumulation.) If the drums are scored, they will have to be machined to restore a smooth surface. If the drums are too worn or deeply scored, replace them. A brake shop or your trailer dealer can test the brake magnets with an ohmmeter to see that they are drawing the proper number of ohms according to the specifications in your trailer owner's manual.

If you have hydraulic brakes, be sure all fluid lines are in good condition and are not leaking.

Brake Check

Several times each year, perform a brake inspection and adjustment. Do this check both with an empty trailer and with a loaded trailer. A dry, hard, level roadway is ideal for this test. Accelerate to 10 miles per hour (16 km/hr) and then brake. It helps to have a knowledgeable observer on the ground to tell you whether a particular wheel is either locking up or rolling free in relation to the others. If so, the noncompliant brake must be adjusted. This involves crawling under the trailer and using a screwdriver-type tool to turn the tensioning screw within the brake drum. If you are not experienced with this, have your trailer dealer do the work for you.

Once all of the trailer's brakes are stopping evenly, proceed with a brake controller test. (Refer to your brake controller manual for specific instructions.) On level ground, accelerate to about 30 miles per hour (48 km/hr). Then, without touching your brake pedal, bring the rig to a stop by using the manual electric brake controller mounted on or under the dashboard. If the trailer brakes either grab suddenly with a lurch or can't stop the rig, adjust the controller per manual instructions. Once the trailer brakes alone stop the rig properly, test by using your brake pedal, which activates the towing vehicle brakes and the trailer brakes simultaneously. The trailer brakes should not lock up on dry pavement when you use the brake pedal.

If your brakes or lights don't seem to be making a connection, first be sure the plug is seated securely in the receptacle on the truck. If there still is no juice, examine the plug for dirt, oil, or debris in the holes. To remove water, cobwebs, and dirt, use a cotton swab or toothpick. If necessary, twirl a dab of steel wool (plain, not the soapy kind) onto the end of a toothpick and polish the hole. If the connection seems clean but still you have no lights, get professional help. When the trailer is not in use, use a plug cap or cover the end of the plug with a dry plastic bag fastened with a rubber band.

Emergency Trailer Brake Check

To guarantee that the emergency trailer brake battery is fully charged and operational, use a voltmeter to test it. It should read 12 volts. Another way to test the system is to remove the breakaway pin from the switch (see page 116). This activates the electromagnets and engages the brakes — you might actually hear a *clunk*. Then drive forward and see whether the trailer brakes lock up. If they do, the battery and mechanism are operational. If the trailer were to become separated from the truck, the breakaway pin would pull out of the switch and the brakes would be engaged. However, this test does not tell you whether the battery is fully charged. You need to meter it to be sure. If your trailer brakes don't lock up at all during the test, either your battery needs charging or there is a problem in the wiring or the brake mechanism.

Tires

Rotate and balance tires once a year or every 5,000 to 7,000 miles (8,065 to 11,291 km) to equalize wear. Check for bare patches, bulges, and other defects. If wear is uneven, check axle alignment. Keep all tires inflated to their proper level. Many tires lose 5 to 10 psi (0.4 to 0.7 kg/cm^2) of pressure during winter storage.

Suspension

At least once a year, or every 3,000 to 4,000 miles (4,839 to 6,452 km), grease the springs and shackles. (Because trailers with rubber-torsion suspension don't have springs, they won't require this.) Check the bushings where the shackles are pinned to the spring ends and the frame. Check shock absorbers and replace when necessary. Tighten axle bolts under the trailer twice a year.

Floor

Check the floorboards frequently for rotting, splintering, shrinking, and warping. Replace any boards that are remotely suspicious. When you do the replacement, choose pressure-treated lumber that is clear (no knots) and the same dimension as the existing floor. You may wish to treat the existing floor with a preservative to combat the effects of manure and urine. Use resilient mats with "life" to help absorb road vibrations and shock. Replace mats when they have become excessively worn. Be sure to check the plywood under the mats on ramps for rot.

Sidewalls

The bottom 2 feet (0.6 m) of the sidewalls of your trailer can sustain a lot of abuse from an inexperienced or scrambling horse. If the walls are metal, check for rust. You may wish to install thick mats or ¾-inch (1.9-cm) plywood over the metal walls as added protection for both the horse and your trailer.

4.1 ADDED SIDEWALL PROTECTION
This steel trailer has ¾-inch (1.9-cm) painted plywood 24 inches (61 cm) high added the entire length of the stall. The scrambling continues but the trailer walls are protected.

General Inspection

Ensure that the hitch, safety chains, chest bars, tail bars, dividers, doors, and windows work properly. You should be checking all of these things each time you use your trailer, and fix and repair as necessary. Inspect safety chains for worn links or cracked welds. Especially if you live in a humid climate, you'll need to clean and oil the teeth of the hitch jack stand to facilitate up-and-down movement. Grease or soap the ball (see photo 10.2) frequently to keep it moving freely in the coupler. Lubricate the moving parts of the coupler as necessary.

Check rubber gaskets and molding around windows and doors to be sure they make a complete seal against rain and weather. Replace when necessary. You may need first to remove rust from the area with a putty knife and wire brush.

Use spray lubricant on any hinges, latches, or other moving parts that do not function freely. When they start to squeak or bind up, it is past the time to treat them. How often you need to lubricate hinges depends on the climate; it might be once a month or once a year.

SAFETY TIP

Mounting a license plate so that it hangs past the lower edge of your trailer is very dangerous. The sharp edge could slice a horse's leg. (See safe license mounting in photo 2.19.)

4.2 UNSAFE WIRING
All wires should be enclosed in a rubber sheath from the plug to the under-trailer wiring. Pictured here is an unsafe situation: Exposed wires are at both ends, with duct tape in between.

Exterior Maintenance

Maintain your trailer's exterior for a tidy appearance and to protect your investment.

- Wash the trailer as needed and wax it at least twice per year.
- Depending on the climate and use, steel trailers usually need to be repainted every three to eight years. Sandblasting may be required to remove rust before painting.
- Aluminum trailers need an annual acid bath to keep them looking new. Check the Yellow Pages for commercial truck washes that offer the service to semi tractor/trailers.

Trailer Storage

Store your trailer on level ground with the hitch jack adjusted so that the trailer's weight is balanced between the tongue and the tires. Elevate the front end just a bit so that rain and snow slide off the roof. If possible, park the trailer out of the weather to preserve the paint job, or buy a tight-fitting trailer cover. (The flapping of a loose cover might do more harm than good to the paint job.)

Parking your trailer on concrete will help protect against tire rot.

If you are going to store a trailer for an extended period, jack up each side and place blocks under the axles where the springs attach. This will take weight off the tires. Be sure the trailer is never parked or stored within reach of horses, as they will likely damage it by chewing or rubbing. If you want to park a trailer in a pasture to accustom young horses to it, use an old, safe trailer.

4.3 WHEEL BLOCKS

To keep your parked trailer from rolling, use one or two heavy wheel blocks on each side of the trailer in between the front and rear wheels. If you only block one side, the trailer can pivot and roll off the jack wheel.

4.4 TRAILER WHEEL CHOCK

When parking your trailer on level ground, you can place the wheel of the trailer jack on a flat board at least 12 inches (31 cm) square. This will allow you to move the trailer somewhat in order to position it when hitching up. In reality, trailers are not very easy to move by hand when you want to move them, and they tend to roll off the board when you want them to stay put. However, a board under the wheel *will* prevent the jack from sinking into soft ground or getting frozen in the mud. A wheel chock, which has a concave surface for the wheel to fit in, works just as well for supporting the jack wheel, plus it helps stabilize the trailer and keeps it from moving.

4.5 Don't Do This

If you need to block under a gooseneck jack for additional height, this is not the way to do it. Here the jack is placed on a trailer ramp that is designed for changing flat trailer tires (see chapter 12). The trailer ramp has a concave surface, which prevents the bottom of the jack from sitting flat on it; thus, the jack has tilted under the weight of the trailer. This could damage the jack and cause the trailer to fall.

4.6 Straight-Pull Hitch Lock

To prevent someone from hooking up to your trailer and stealing it, consider using a hitch lock whenever you park. This type of lock works well on a straight-pull trailer. (See also photo 2.6.)

4.7 Gooseneck Hitch Lock

This is an example of a hitch lock for a gooseneck trailer that has an integral lock, so it doesn't require a padlock.

5
✦ Maneuvering Your Rig ✦

Take the time well before a trip to become proficient at maneuvering your truck and trailer. If you have a level field at home, practice maneuvers using cones or plastic barrels or buckets. When you feel confident with your skills at home, take the empty trailer to a parking lot. During the week, church parking lots are often empty; they make good level "playing fields" for your trucker-school drills. Eventually, put pressure on yourself and maneuver between cones in a parking lot during business hours. You'll be glad you did when you have to park your trailer in a tight space at a horse show or the vet's clinic. It is much better to crunch a few cones in the grocery store parking lot than it would be to smash into the next trailer at a show.

If you are new to trailering and are hesitant to take on the responsibility, don't worry too much. Confidence comes with experience. Go along with a competent, seasoned hauler to get a feel for the whole thing. When it comes time for you to take either practice trips or real ones, always ask someone go along with you. But choose your companion carefully: You need someone you can rely on to assist you in case of emergency. Sometimes having the wrong person along can get you flustered and undermine your confidence. Having the right traveling companion gives you a great feeling of teamwork and confidence, which usually means a safer trip.

Learn to Use Your Mirrors

Because you will often be accelerating and traveling at a slower speed than passenger cars, always be on the alert for someone coming up behind who wants to pass. Adjust your mirrors so you can see alongside and behind the trailer (photo 5.1). When passing, use the mirrors to ascertain that the trailer has cleared the vehicle you passed before you start pulling back in the lane in front of it.

When an assistant is helping you to back up or maneuver the trailer, it is his responsibility to be sure that he can see you and you can see him in your side mirror at all times (photo 5.2). It is very difficult for a driver to navigate safely when turning around to look for a helper.

5.1 While you are driving, use your side-view mirrors to check behind for traffic or for doors that might have come unlatched.

5.2 Make specific signal arrangements ahead of time to avoid a crunch or divorce court!

Backing Up a Trailer

Do you have back-lexia when it comes to maneuvering your trailer in reverse?

When you are backing up a trailer, to get the trailer to go to the left, the rear end of your truck should go to the right, which means you turn the truck's wheels and the steering wheel to the right (clockwise). You do this while looking in a mirror, which makes your brain do another left–right switch. Although this becomes second nature when you've been trailering a while, when you first start it can be very confusing.

First of all, take your time. There is never any hurry, and especially when you are backing up. Take it slow. Get out of the truck, walk around, and take a look at where you are trying to place your trailer. That often gives you a visual goal and you can see how much space you have and what there is that you might hit.

Second, if the mirrors are really confusing to you, turn around and look out the window — this is no big deal. Finally, if you are still having trouble knowing which way to turn the steering wheel, use the easy tips that follow.

Backing Tips

When backing up a trailer, an easy way to remember which way to turn the steering wheel is to put your hand in the center of the bottom of the wheel. If you want the back of the trailer to go to the left, move your hand to the left. To get the back of the trailer to go to the right, move your hand to the right.

For a sharp turn, turn the steering wheel before you press the accelerator. For a gradual turn, turn the steering wheel and press the accelerator at the same time. Always back up slowly. Once the trailer is going in the direction you want, straighten out the truck wheels so the truck can follow the trailer.

5.3 Cone Alley
On your first trip to the parking-lot practice school, set up a situation that requires you to back straight into an alley of cones. Alternatively, use the painted lines on the parking lot as a guide.

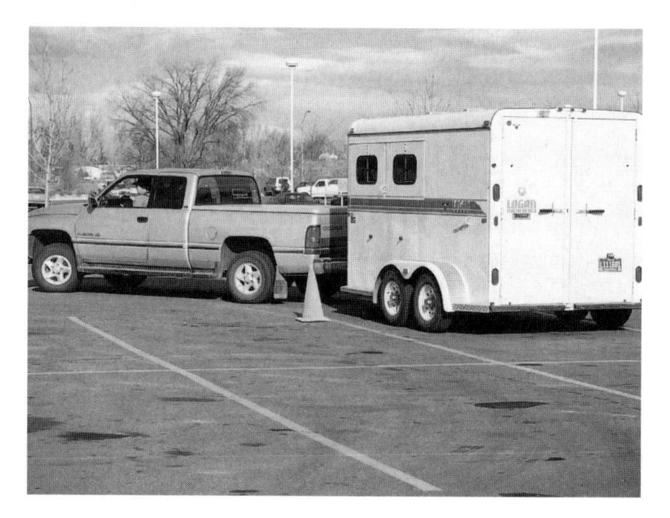

5.4 Turn Left around Cone
Set out a single cone and use it to become familiar with the turning radius of your rig. First, turn to the left and keep a close eye on the cone. Remember, the longer your rig, the more the rear wheels of your trailer will cut inside the track of the truck. Because goosenecks are hinged so far forward on the truck, the trailer tends to "cut the corner" and so requires a wider turn than does a straight-pull.

5.5 Turn Right around Cone

Now use a single cone to practice turning right. Whoops! This is not as easy because the cone is more difficult to keep track of in the right mirror. When you lose sight of the cone, the next time you see it, it might be crunched!

5.6 Turn while Backing

Next, return to the backing exercises, this time incorporating turning as well. Whenever possible, set it up so that you are backing with the bend in your rig on the driver's side, where you can see what is happening. However, eventually you will have to learn to back from the right for places where you are backing the trailer into a blind spot. Start out giving yourself plenty of room and make a large, sweeping turn backward until you get lined up. Alternating looking in your mirror and out the back window may help, but it could confuse you. (See chapter 13 to learn about parking at horse shows.)

If you find that you've turned too late and have overshot your target, pull ahead to straighten out, then try again. When making an adjustment, don't overcompensate. If you begin jackknifing (that is, when you make such a sharp angle that the trailer might crunch the back corner of the truck), stop immediately.

Driving Tips

Be considerate of your horse as you drive. Although it is illegal for you to ride inside a tag-along trailer on the highway, you can see what it's like to ride in the trailer by taking a short sensory-awareness trip around your own property.

- Drive safely. Never go over the speed limit, and don't be reluctant to drive under it. Let people pass you.
- Leave room to stop. Realize that when pulling a trailer, it requires a longer distance for you to stop your rig.
- Consider your horse's comfort. Make all acceleration and deceleration gradual.
- Take curves at moderate speed to prevent scrambling and swaying.
- Slow down for obvious bumps or holes.
- Use turn signals well in advance.
- When taking a corner, wait until your vehicle is completely straight before you accelerate; otherwise, your horses may lose their balance.
- Anticipate stop signs and corners, so you have ample distance for slowing or stopping.
- Don't make any quick lane changes.
- Keep your senses alert to unusual sounds, smells, and vehicle motions.
- Keep your radio and CD player off so you can hear suspicious sounds.
- Check your mirrors often for open doors, signs of smoke, or other unusual events.

Swaying or Fishtailing

It's likely that at some time during one of your trips you'll feel the trailer swaying. Swaying is more common with a straight-pull trailer than with a gooseneck. The motion can be caused by a flat tire; low tires on truck or trailer; an unbalanced load (usually the trailer is loaded too heavily toward the rear, resulting in too little tongue weight, or you are hauling horses facing the rear in a trailer not designed for it); a too heavy trailer; going too fast (especially downhill); moving horses; wind gusts or turbulence from passing trucks; an irregular or washboard road; poor wheel alignment or balance. Control gentle swaying simply by keeping the tow vehicle straight, letting up on the accelerator, and slowing down. If your trailer sways regularly, stop and check the tires and make sure your horses aren't scrambling around. Balance the load if necessary by moving items from a rear trailer tack room, for instance, to the front tack room. Minimize trailer sway by making sure that the truck is always "pulling" the trailer, that the trailer never "pushes" the truck. If the trailer gets too "pushy," slow down the trailer and keep the truck moving forward. Sometimes the very thing that will straighten out a trailer sway is to accelerate and get the truck positively *pulling* the trailer again.

Violent swaying or "fishtailing" can be very frightening and unnerving, but don't panic. Keep your eyes on the road, don't try to correct via the steering wheel, and *don't hit the brakes*. If you slam on the brakes, the trailer can jackknife and cause a serious wreck. Keep calm, continue straight ahead, and gently apply the manual trailer brake mounted under the dashboard while gradually letting up on the accelerator. This will diminish the swaying and allow you to maintain control of the rig. Practice this exercise ahead of time — keep the truck moving straight ahead, reach for the trailer brake controller, gradually let up on the accelerator — with an empty trailer until you feel you have mastered the reflex urge to put your foot on the brake pedal.

When you do apply the truck brakes, do so gradually (with even pressure if you have antilock brakes or with a slight pumping action if you don't have antilock brakes). Never slam on the truck brakes unless an emergency situation leaves you no alternative, because when you slam on the truck brakes, the trailer is likely to jackknife.

5.7 Slow down for curves. Some curve signs have a speed limit posted, but don't wait for a sign to tell you to slow down when you will be twisting up or down a road. You have three methods to slow your rig: vehicle brakes, trailer brakes, and downshifting the transmission.

5.8 Use lower gears when going downhill. In a lower gear, the engine helps to slow the vehicle and takes the load off the brakes, which can prevent your brakes from overheating. Some hills have a low-gear reminder, but many (such as in photo 5.7) don't. Again, take the responsibility to drive according to your load and the contours of the road.

PART II
TRAINING

Training a horse to load in a trailer is no different from any other aspect of horse training. Start with the basics and build using a progression of lessons. It helps to have a clear picture of your end goal in mind, but to get there you must develop a group of smaller subgoals. Treat each subgoal as a separate lesson. If you and your horse master the ground-training lessons that follow, your horse *will* load in a horse trailer.

As you go through the lessons, show the horse what you *want* him to do and what he must *not* do. Horses are much more content when they know absolutely what is expected of them. Remember, horses are basically followers and will do almost anything for you when your requests are clear, consistent, and fair.

Consistency

Let your horse know each time he has made a mistake. If you don't, it will be confusing for him and it will take him longer to learn the correct response. For example, each time he puts pressure on the halter, whether he is just lightly leaning on it or trying to blast past you, give a tug on the halter to let him know he should not do that. Your goal is to have him lead lightly — like a butterfly on a string, not a runaway freight train.

Praise

The doling out of praise is a little bit different. At first, praise your horse each time he responds correctly. You can lavish the praise on him with a good rub on the withers or stroke down his neck as you say "Good boy" or "Good girl." In later lessons, eliminate the scratch or stroke but continue with the verbal praise. Finally, when the horse knows the lesson thoroughly, eliminate the verbal praise. Then you can use the verbal praise and/or the scratch or stroke occasionally to reinforce his good habits. Once the lesson has been learned, praise is most effective when it is used sporadically. If you lavish praise continually, it will be hard for the horse to distinguish exactly what he is being praised for and the reward will lose its effectiveness.

Whether you are correcting your horse or praising him, your action should follow the behavior immediately. If you are slow to respond, you may be punishing or rewarding the next behavior instead!

Repetition

Once your horse has learned a particular lesson, repeat the exercise regularly over a period of days and in different locations to establish it firmly in his mind. Repetition is the key to developing a conditioned response, which is a requirement for solid horse training. Don't think because you can load your horse on a warm, quiet day when his buddy is already in the trailer that he knows the lesson definitively. Can you load him when the wind is blowing, his buddy is in the barn calling to him, and a dog is barking nearby? The more thorough you are with the entire ground-training program, the more assured you will be of your ability to load your horse during exciting times such as during a storm, along a busy highway, or when he is injured or ill.

Cooperation

When you establish a solid base of in-hand work and work over obstacles, leading or sending your horse into a horse trailer will be easy. Trailer loading should not be a battle between human and horse. In order for the human to win, it is not necessary for the horse to lose. A horse should not be mechanically or physically forced into a trailer. He should enter willingly, of his own accord. That way, both you and your horse are winners.

Physical and Verbal Aids

As with other training lessons, your mind is a powerful aid in guiding the horse. But you also need physical aids to tell him what you want. These aids are your body language, a halter and lead rope, your hands, an in-hand whip, and, in some cases, a chain for the halter. As the horse learns what you want him to do, start coupling verbal commands or signals with the physical aids so the horse doesn't depend on the physical cues.

For example, in one of the simplest lessons, "Walk on," when leading the horse, you want him to walk forward when you walk forward. At first you might need to use a tap with the whip or a pop with the end of the lead rope on the horse's hindquarters to cause him to step forward promptly. Or you might need a sharp tug or two on the halter or chain to keep him from charging forward or crossing in front of you. Or perhaps you might need to poke him in the neck or shoulder with your elbow or the butt of the whip to keep him from crowding you.

But once he has learned to respect your personal space and walk forward promptly with you, subtle voice commands, sounds, or gestures will confirm the guidelines for the horse. Then, later, he will just operate from your body language or any specific cues you want to use. Remember, as you progress through this battery of lessons, it is better to perform the simple lessons well than to rush ahead to the end goal ill prepared. Do your homework, take your time, and be patient.

The Cast of Characters

For the training examples in this book, I have used one very seasoned loader and traveler to demonstrate some of the ideals you are aiming for. Except for that mare, the other horses in the photographs are either learning their lessons for the first time or having a review. Here is the cast.

- **Veteran:** A 25-year-old Quarter Horse mare that I've had since she was a weanling. She's the experienced, dependable horse on the cover that I describe in the preface.
- **Mr. Mellow:** A very people-oriented, lovable Selle Français/Quarter Horse gelding that is very tractable but a little bit slow (cold-blooded) in his responses. Because of his tendency to nuzzle and invade space, he could easily become a large, spoiled pet. He's very curious and likes to inspect everything with his muzzle.
- **Ms. Antsy Pants:** A Trakehner/Quarter Horse filly that is a bit impatient. When asked to do something new, she'd never refuse but she might try to hurry or push.
- **Reflex Queen:** A Trakehner/Quarter Horse filly that has unusually "large" reactions to physical stimuli.
- **Rookie:** A 12-month-old Quarter Horse filly with very minimal in-hand training. She's built like a tank and has a strong mind. She can be tough when first learning certain new things but is confident and solid when she has mastered them.

Overview

In chapter 6, you'll learn important in-hand lessons that are the basis of all horse training and handling, whether trailer loading, longeing, riding, or just grooming and health-care procedures.

Chapter 7 provides ideas for obstacle training. Obstacles are objects that your horse negotiates in a particular fashion to build his confidence and to develop skills that will be useful during trailer loading.

In chapter 8, I'll demonstrate loading and unloading in a number of ways, using horses at different stages of training and with various style trailers.

Because of the progressive nature of the program, you won't see too much misbehaving, but in chapter 9 I'll point out common trouble spots and how to avoid or correct them.

6

✦ In-Hand Work ✦

Both you and your horse must be comfortable working with each other during in-hand work. Before you even think about attempting to load your horse in a trailer, be sure you have established in your horse thorough in-hand manners and responses. This chapter outlines the in-hand work that every horse should know.

Preloading Ground-Training Checklist

You should be able to perform all of the following with your horse relaxed, cooperative, and with no resistance. All exercises should be performed from both the near and the off side.

- ❑ Catch in stall, pen, paddock, and pasture
- ❑ Halter without fussing
- ❑ Stand tied without pulling, pawing, swerving, or chewing
- ❑ Unhalter without pulling away
- ❑ Turn loose without galloping away
- ❑ Walk forward promptly and in proper position
- ❑ Turn left from light cues
- ❑ Turn right from light cues
- ❑ Stop without requiring strong cues
- ❑ Back without requiring strong cues
- ❑ Stand on a long line without moving while trainer moves around
- ❑ Turn on the forehand from each side
- ❑ Move sideways from light cues

6.1 Stand Tied

From a very early age, all horses must learn to stand tied without pulling, swerving, pawing, or chewing. The best way for a horse to learn to stand patiently while tied is to begin with a short session of 5 minutes and gradually increase over a period of weeks until the horse will stand quietly for several hours. This is an essential prerequisite to trailer loading, traveling, and standing tied to the outside of the trailer. If a horse does not know how to stand tied to a hitching post or rail, when he is tied inside a trailer it is likely he will pull back or try to rear when he feels the pressure from the halter. Be sure your horse knows the standing-tied lesson very well before you attempt to load and tie him in a trailer.

Safe, Strong Halter and Lead Rope

Web Halter

The advantages of a web halter with a snap-on lead rope (compared to a rope halter, which is discussed next) are that the lead rope can be removed quickly and easily, the surfaces of the halter are wider and milder, and you can add a chain if necessary. One disadvantage is that the rig is only as strong as the hardware on the halter and the snap and attachment of the lead rope, which is always a consideration when tying a horse. More importantly, however, some horses don't respond to the pressure from the mild, wide webbing and become pushy and headstrong.

Rope Halter

The advantages of a rope halter (compared to a web halter and snap-on lead rope) are superior strength and a narrower surface of communication with the horse's poll, nose, and throatlatch, which results in a better response in some horses. When a horse is backing up and you don't want him to, you can deliver more effective poll and throatlatch pressure with a rope halter than a web halter to stop the behavior. The disadvantages of rope halters are that they are a little more difficult for a novice to properly fit and fasten, and the lead rope is more difficult to remove quickly.

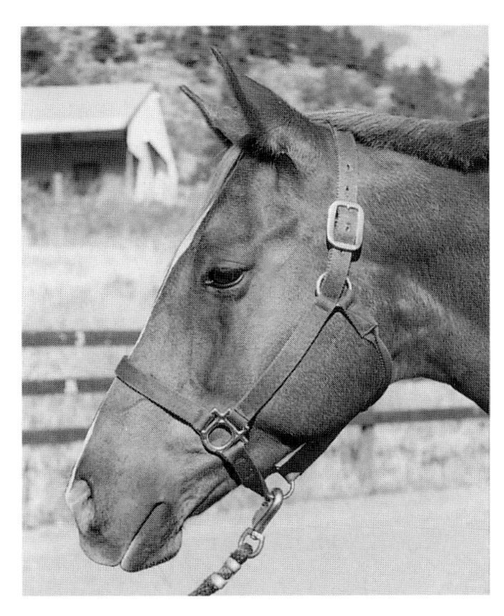

6.2 Web Halter
This shows a correctly fitted web halter and lead rope. This average adult-horse-size halter (for 800- to 1,100-pound [363- to 499-kg] horses) fits this gelding correctly when buckled on the middle hole. The noseband rests two fingers below the prominent cheekbones, the crownpiece rests just behind the ears, and the halter's throatlatch fits in the gelding's throatlatch. Snapped to the halter is a 10-foot (3-m) lead rope with a bull snap on the end.

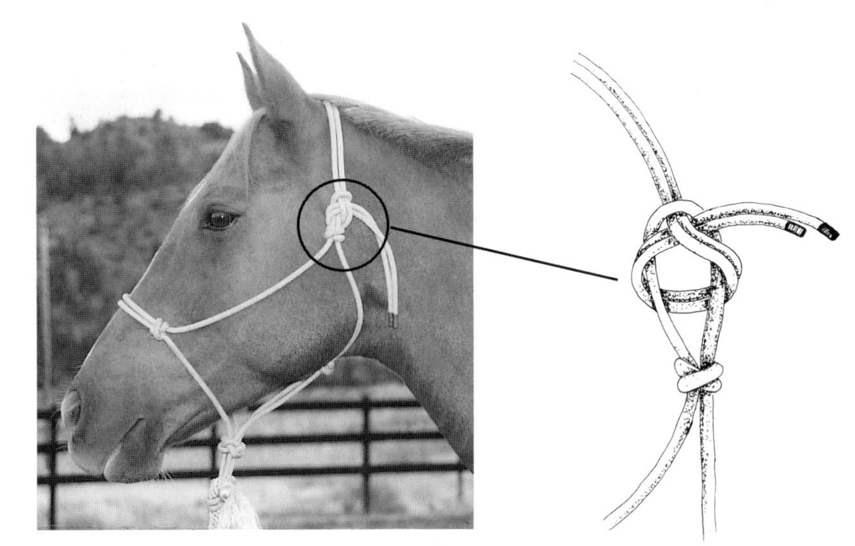

6.3 Rope Halter
Here you see a correctly fitted rope halter (Double Diamond, Gallatin, MT). This average-adult-size halter fits the same gelding. The throatlatch is positioned in the horse's throatlatch and the knots at the noseband are two fingers below the prominent cheekbones. The rope halter is fastened at the near cheek with a sheet-bend knot (see illustration). The halter is made from one continuous piece of rope and the lead rope is attached to the halter via a knot. These halters have rarely been known to break.

Sheet-Bend Knot
Use the sheet-bend knot to fasten a rope halter.

Safe, Strong Halter and Lead Rope (continued)

6.4 Web Halter with Chain
If your horse is large or headstrong, you can use a rope halter (see photo 6.3) or add a chain to a web halter. Use an 18-inch (46-cm) stud chain. Thread the snap at the end of the chain through the near-cheek ring of the halter, over the noseband, under the noseband, through the off-cheek ring, then up to the off-round ring where the snap is fastened. Snap the lead rope to the triangle at the end of the chain. This chain is in an "engaged" position.

6.5 "Parked" Chain
Be prepared by having the chain in place but not engaged. You can "park" a chain on a halter so that if you need it, you have it already "installed." Give your horse the benefit of the doubt. Until you do require the chain, snap the lead rope to both the chain and the throat ring of the halter. This "parks" the chain, where it will be inactive until you deem it necessary to use. At that time, just unsnap it from the halter ring.

Head Down

Teaching your horse to put his head down instructs him to obey your request to respond to pressure. More importantly, it relaxes and centers your horse.

6.6 Tall Horse, Head Up
Notice the startled look in Mr. Mellow's eye. He is concerned about things elsewhere. With his head this high and his neck rigid, it would be hard to handle or lead him. With the lead rope looped over my left forearm (not coiled around, which would be dangerous), I prepare to calm him.

6.7 Head Down

With my left land on the bridge of his nose and the fingers of my right hand at his poll, I exert very light pressure and he immediately lowers his head. Notice the calm, relaxed expression. This gelding has long ago learned this lesson, so it is a handy tool to help "center" him when he is distracted.

6.8 Center Horse's Head

To teach a horse the head-down lesson, use a rope halter or a chain on a web halter. Rookie has had limited handling and isn't yet convinced that she should pay attention. When I place my right hand on her poll, she moves her head away but not her body. I reach for the lead rope so I can reposition her head in front of her body.

6.9 Position Hands

Now that Rookie's head is in front of her body, she is already more focused. With slack in the lead rope and the rope draped (*not* coiled) safely over my left forearm, I place my hands in the head-down position. Some horses must first be taught to accept touching. A horse that is not accustomed to having her head handled may try to shake your hands off her face. This filly is calmly accepting, so I can move to the next stage, which is exerting pressure.

6.10 Slight Poll Pressure

It requires only the smallest amount of pressure to show Rookie what I want. She has a very peaceful expression and hasn't moved one foot during the whole exercise. If a horse does not know "Whoa" on a long line, you must teach that before "Head down."

“Whoa” on a Long Line

“Whoa” on a long line is the equine equivalent of putting a horse on the honor system. You are teaching and testing a horse’s ground training to be sure that when you tell him “Whoa” and leave him, he stands perfectly still until you release him. It is like the command “Stay” for a dog and it is prelude to “Ground tie,” which you might want to use when riding. When you first teach a horse this lesson, you will step away from him only about 1 or 2 feet (0.3–0.6 m) extra and for only a few moments at a time. Gradually increase the distance and time until you can leave your horse and step to the end of your 12-foot (4-m) lead rope for several minutes.

6.11 Develop Patience

Ms. Antsy Pants derives great benefits from this lesson. It helps her focus and develop patience. It is not important to square up the horse’s legs as if you were in a conformation class. However, it’s a good idea to leave the horse in a balanced, comfortable position because she is then more apt to stand still. Always return to your horse. Never ask the horse to come to you at the end of the lesson. Return to the normal in-hand leading position and “release” the horse from the “Whoa” command by saying “Okay” or “Walk on” or whatever is appropriate for what you want to do next.

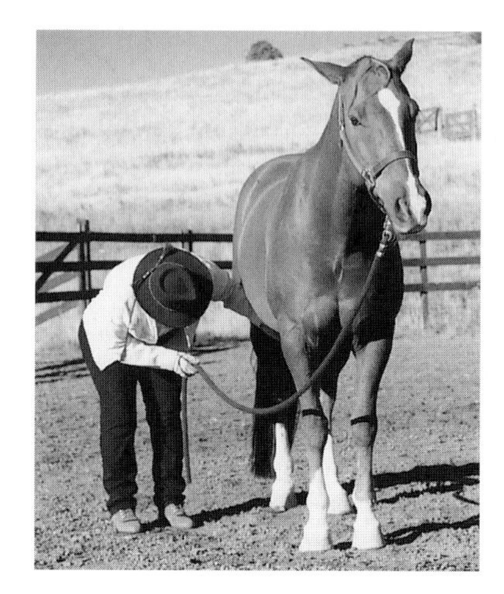

6.12 Test

The “whoa” on the long line lesson extends to situations where you not only leave your horse but you also do something “suspicious” where he cannot see you. For example, here I’m checking the area in front of Mr. Mellow’s sheath for fly bites. You can tell by his expression that he’s wondering what I’m doing, but he has learned “Whoa” on the long line, so he doesn’t move away. This will prove handy when you are loading a horse, for example, and you need to make an adjustment to his blanket strap or a hind traveling boot.

6.13 Final Exam

Finally, test the thoroughness of the “Whoa” lesson by draping the lead rope over the horse’s neck and leaving him. At first hold this lesson in an arena or other fenced area. Lessons that develop patience in your horse will pay large dividends in all work to come.

Leading — Demonstration of Goal

6.14 In-Hand Position
Work at your horse's shoulder with about 2 feet (0.6 m) of slack lead rope between her halter and your hand. This position is safe — you won't get stepped on or left behind. And it is an effective position to get the horse started, control her forward movement, and shape turns.

6.15 Walk On
When you take the first step forward, do so energetically. Your body language encourages the horse to respond in kind. Notice the mirror imaging of my legs and Veteran's. This is a desirable, prompt, energetic forward step.

6.16 Trot
If your horse needs "waking up," trot from the halt. Give your horse more headroom (about 3 feet [1 m]) and use the voice command "Trot on." You want to develop an instant, energetic reaction from your horse. Practice walk and trot departs regularly so they become second nature to your horse. Then, if a horse has a question when asked to step into a trailer for the first time, the practice will pay off. The horse's forward response to your cues will be so deeply ingrained that she will automatically step forward.

6.17 Walk on the Long Line
When your horse knows "Whoa on the long line" and the "Walk" command, teach her to walk forward on the end of a long line. You just want her to walk straight forward a few steps, not turn in toward you as if on a circle. My right hand is signaling Veteran to step forward at the same time I say "Walk on." Once your horse starts moving, you will either walk along with her or ask her to whoa. The object of this lesson is to know that you can start and stop your horse from 8 to 10 feet (2 to 3 m) away. Work on these basic lessons until your horse has mastered them.

Lagging and Correction

A horse that is not paying attention is not likely to be in tune with your body language and your commands. A distracted horse will tend to be a slow-starter and lagger during in-hand work. Discourage lagging.

6.18 Lack of Focus
Mr. Mellow is sleepy and paying attention to a pasture mate. He's relaxed but he's not focused. Let's see what happens when I walk away at my normal pace.

6.19 Left Behind
He's very late starting and within a few steps I'm way ahead of Mr. Mellow. I'm in a dangerous position and have no way to influence his impulsion or direction. This is a very bad in-hand response and an unproductive leading position.

Never alter your pace to match a slow horse. Instead, wake the horse up. Start in position at his shoulder. Carry a 30-inch (76-cm) dressage whip in your left hand and at the instant you say "Walk on" and start walking yourself, tap the horse on the hindquarters with the whip. It should be a controlled, subtle maneuver, not a wild gesticulation. Do this every time you walk off with the lazy horse and soon he will wake up and form a new habit.

6.20 Tackless Test
In your in-hand work, imagine you are using invisible tack. This should make you less dependent on controlling your horse with tack and make you focus more on body language, voice commands, and consistency. Start in the normal in-hand position, then remove the halter and hold it in your left hand.

6.21 Walk On
When *you* are ready, step forward just as you do when your horse is haltered. Use your normal voice command. At first you can hold your right hand as you would if holding a lead rope as a visual cue to your horse. Later, leave your right arm at your side, look straight ahead, and walk straight ahead.

6.22 Ideal Relationship

To me, Veteran has an ideal attitude. She's responsive and cooperative, yet she's a confident individual who knows her role. She is a horse! I like that! She respects my personal space whether we are working or standing. This is the type of relationship you want to develop with your horse.

6.23 Don't Crowd Me!

Some horses, like Mr. Mellow, when given the opportunity, will crowd their handlers. This can stem from any number of reasons. The most common are insecurity, pushiness, and, as in this case, excessive sociability. This gelding gravitates to all humans, other horses, and animals and solicits attention and touching. He would love to be someone's "pet." Here, as is shown by the slack lead rope and his relaxed expression, he has come to a stop very close to me and then curled his head to come even closer. While this might seem darling, it can be a nuisance and even a hazard.

One way to encourage a horse to stay in his own space and keep out of yours is to use the butt end of your in-hand whip as a cuing device. I prefer to use a 30-inch (76-cm) dressage whip with a mushroom-cap end to push on the horse's shoulder or neck. If a horse is particularly stubborn, you can use a whip without a broad cap to deliver more of a poke.

6.24 Move Over!
At first it might take quite a forceful press on the horse's shoulder to get his attention and start him thinking about moving a step away from you.

6.25 Better Response
After just a few deliberate cues, all that is required is a light tap and Mr. Mellow steps wide to the right. This personal-space lesson will come in handy when you teach your horse in-hand turning and side passing.

6.26 Elbow Tap
Another technique to move the horse's body away is to use your right elbow to poke him over. Yes, I said poke. Use your elbow in a tap-tap manner. Don't lean into the horse — he'll just lean right back.

6.27 Pitch a Wave
If the horse is just crowding you with his head, pitch a wave in the lead rope that moves his head away from you.

6.28 Success!
Ahh, this is much better. Now we're ready to start doing something productive.

Turn on the Forehand

A turn on the forehand is one where the horse rotates the hindquarters around the forehand. If you are standing on the near side of the horse and you ask the horse to perform a turn on the forehand, you are asking him to step to the right with his hind legs. You want him to cross his left hind in front of and past the right hind and then uncross the right hind from behind. Meanwhile, the horse is swiveling on the left front foot, which is the pivot point of the turn, while taking tiny steps to the right and forward with the right front foot. Depending on the trailer, you might use the turn on the forehand or a variation of it to turn your horse around in the trailer or to line him up for loading.

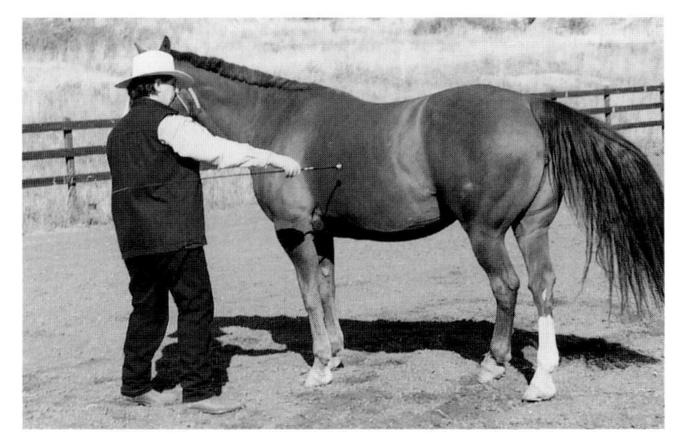

6.29 Set Up Your Horse

To set up a horse for a turn on the forehand left (as described above), tip the horse's nose slightly to the left and cue the horse on the ribs. Here you see the butt end of the whip at the rib area, and Veteran has already stepped to the right with her left hind for the first step of the turn.

6.30 Deep Crossover

In this turn, Veteran is crossing over very deeply with her left hind. Her left front hoof looks twisted because it faces the same direction it was when she started the turn! As she uncrosses her right hind and steps right, she'll pick up her left front leg momentarily to reorient it in the new direction.

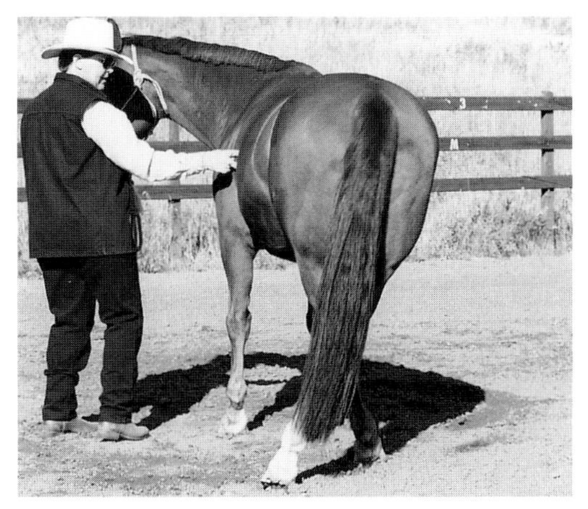

6.31 Lose the Whip

Now perform the turn without the whip and use only your fingers to give the physical signal.

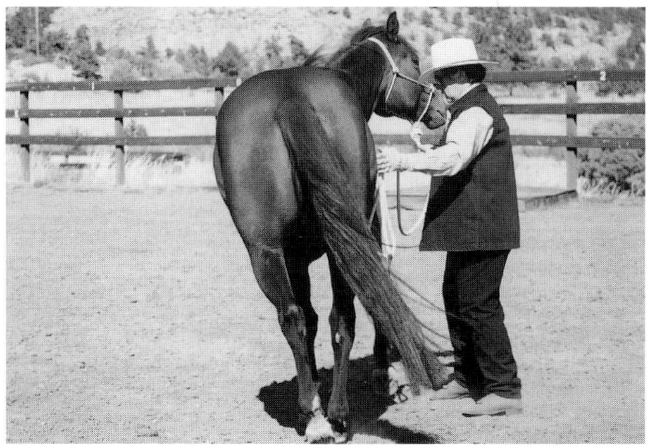

6.32 Work from Off Side

Perform all in-hand work from the off side, too. In a turn on the forehand right, Ms. Antsy Pants' nose is tipped to the right. My fingertip pressure on her ribs is causing her hindquarters to move to the left. She is ready to step to the left with her left hind leg.

Side Pass

When side passing, the horse is walking sideways and his feet move in the same footfall pattern as the forward walk: left hind, left front, right hind, right front. In a side pass to the right:

1. The horse picks up the left hind and crosses over in front of and past the right hind.
2. Then the left front crosses in front of and past the right front.
3. The right hind uncrosses from behind the left hind and steps to the right.
4. The right front uncrosses from behind the left front and steps to the right.

Note: Steps 2 and 3 happen almost in unison, allowing the horse to maintain his balance.

The side pass will come in handy in several ways for trailer loading. You can use it to position a horse to enter a trailer if he has approached it off center or tried to avoid it by sidestepping one way or the other. It is also helpful to position him in a slant-load or stock trailer once he's inside. The cues you use along his ribs will also be handy to keep his body straight when he learns to back off an obstacle or out of a trailer.

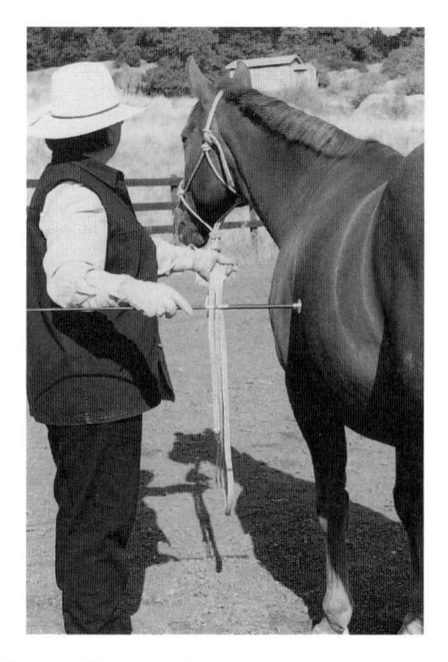

6.33 Side Pass Setup
For a side pass right, stand alongside your horse facing her left side. Hold the lead rope in your left hand a few inches from the halter so you can keep the horse's head and neck fairly straight, with a slight flex to the left as shown here. Hold your whip in your right hand, with the butt of the whip aimed at the horse's ribs at the approximate position your leg would be if you were riding.

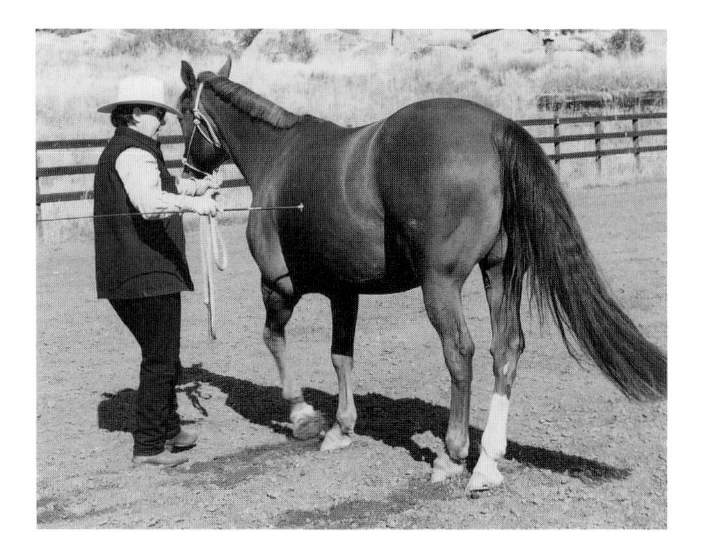

6.34 Keep Horse Straight
Walk toward your horse keeping your aids consistent so that the horse's body stays straight. For each step sideways that you want, press the whip butt on the horse's ribs. Use a voice command such as "Oh-ver." Here Veteran has just crossed the left hind over the right hind and is now crossing the left front over the right front. She's in the process of picking up her right hind and uncrossing it from behind the left hind. Notice that the aids are consistent and the horse's body remains straight.

Next, eliminate the whip and use just your fingers to cue your horse. *Note:* Now that you no longer have a whip to reach your horse, you have to step closer to her to reach her with your fingertips.

6.35 Voice Only

Now practice the side pass with just the voice command. Here I'm straight and Veteran is straight, but her head and neck shouldn't be tipped quite so much to the left. With this trained horse, the excessive left bend is okay, but in an untrained horse it could cause the horse to turn on the forehand instead. As with all exercises, practice the entire side pass sequence from the off side: with whip butt, with fingertips, and finally with only the verbal command.

You can use the end of your lead rope instead of a whip to cue your horse for movement. However, the dual role of the lead rope might be confusing for some horses. Certain horses have a hard time distinguishing the difference between a lead rope that is cueing for movement and a lead rope that is just being held by the trainer for leading. They tend to keep an eye on the lead rope and react to its movement at times when you don't necessarily want them to. Since an in-hand whip is used strictly as a pressure cue for movement or as a visual cue, it is usually quite clear to a horse how to react to a whip cue.

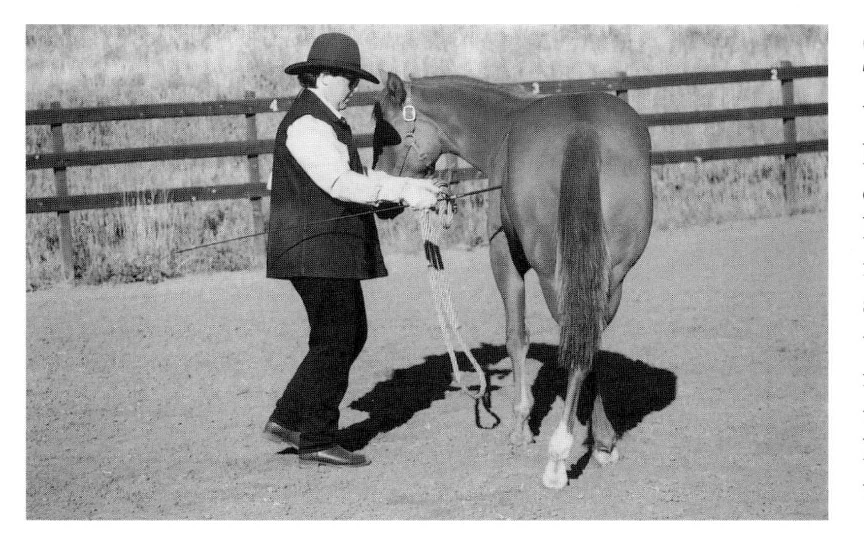

6.36 Review Forehand Turn

When teaching the side pass to a horse, start by reviewing the turn on the forehand. Here Rookie has crossed the left hind over the right hind nicely. Her weight is solidly on the pivot point of the turn, the left front. Her head and neck are turned to the left, and the whip butt is being applied to her heart girth area, causing her to rotate on her left front leg.

6.37 Hold Trainee Straighter

Now the aids are changed for the side pass. Rookie's head and neck are held straight and the whip butt is moved rearward about 10 inches (25 cm) because I want her to move her whole body sideways, not just her hindquarters. At first, I don't expect a huge lateral movement, just a response.

Once Rookie has the idea that I want a forward and sideways movement, I tip her neck and nose slightly to the left but keep her body relatively straight. She then assumes the correct position of her practiced herdmate earlier in this section. At first just ask for a step or two and give lots of praise. Practice often and from both sides.

6.38 Restore Forward Element

If your horse starts to back up instead of moving sideways, release the sideways driving aids, reestablish forward movement, and then return to the side pass. Here I am tapping Mr. Mellow lightly with the whip on the top of his croup to keep him thinking "forward and sideways."

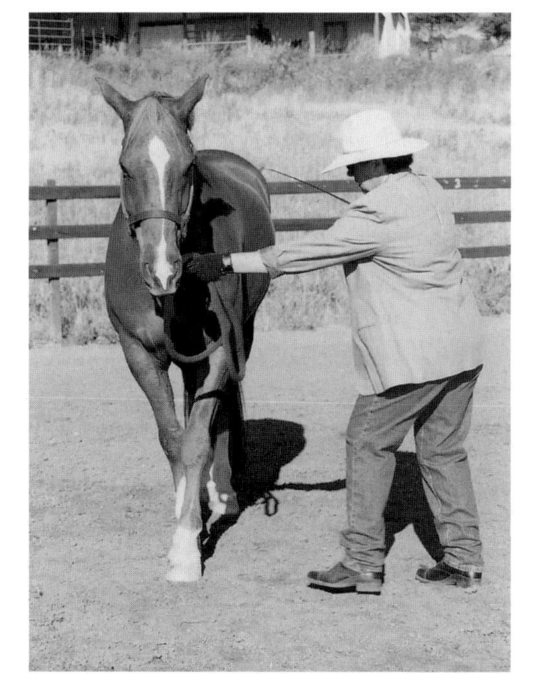

6.39 GOAL
The back is a two-beat diagonal gait in reverse. The left front and right hind lift, move backward, and land together, followed by the right front and left hind. Performed correctly, the horse's body must be straight, the horse relaxed, and the neck low. If the horse's head is raised, it tenses and hollows his back, making leg movement difficult. Veteran demonstrates ideal form and energy for the back.

6.40 SHOULDER POINT PRESSURE
When training a horse to back, face the horse and hold the halter rope about 4 inches (10 cm) from the halter with your left hand. Press your right fingers on the point of the horse's shoulder. Here Ms. Antsy Pants has broken the diagonal pairs, turning the back into a four-beat gait. Although four-beat backing is used when a horse picks his way slowly and carefully through an obstacle, it is not a true back. Aim for a crisp, prompt, balanced two-beat back.

6.41 USE RAILS
To teach your horse to back straight, use ground poles, railroad ties, or hay bales to form an alley. Here 10-foot (3-m) railroad ties are set 18 inches (46 cm) apart. Lead your horse forward through the alley, then stop, turn around and face him (switching hands on the lead rope), and prepare to back. Be sure you keep your horse's head and neck low and relaxed.

Because you walked your horse into the alley, he has a memory pattern of what he is backing through. Most horses have good proprioceptive sense — the ability to move their legs and body in a coordinated fashion without being able to see where they are actually stepping.

If your horse's hindquarters swing to the right, you need to move your left hand under his chin to the right to straighten him. When backing, use the horse's head like a rudder to steer him. If you were to move your hand to the left, it would cause his forehand to come toward you and his hindquarters to swing off to the right even more.

6.42 Tackless Back

To test whether Veteran is dependent on the halter to back her and keep her straight, I try backing her without a halter. In an enclosed pen, I remove her halter and face her as I do for normal backing. I check to see if she keeps her body straight, her topline relaxed, and her head low. Yes. But because of her extreme relaxation, her front is landing slightly in advance of her hind, which means she is a bit heavier on the forehand. I should perk her up with a *click-click* sound.

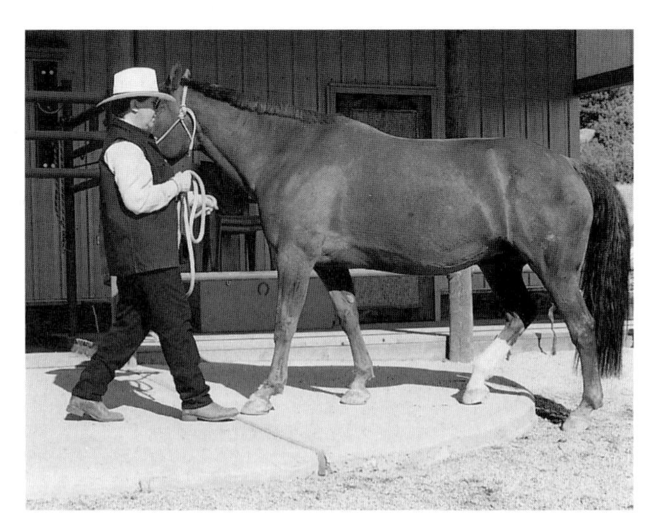

6.43 Back and Step Down

The final backing exercise is the prelude to trailer unloading. Find a small step such as this 4-inch (10-cm) step up onto a concrete slab. Walk your horse onto the slab. Then stop, turn around, and change hands on the lead rope. Back the horse off the slab, taking care to keep the horse's body straight. Whenever a horse steps down with his hind legs, his head and neck automatically come up to balance. (See more of this in photos 7.24 and 8.38.) This exercise is a beginning to help you and your horse get used to the step-down process. Here, Veteran's head elevation is mild because the step down is low. When you ask a horse to back and step down at the same time, the back is no longer a two-beat diagonal gait. The horse must move each foot one at a time.

6.44 Follow-Through

Continue backing and follow-through, keeping your aids and the horse's response consistent: the rhythm, your position, the horse's relaxation, and the horse's straightness.

7

✦ Obstacle Training ✦

A horse should become accustomed ahead of time to several new sensations that he will experience during trailer loading and unloading:

- Low light inside the trailer
- Step up
- Hollow sound of the floor
- Presence of the ceiling over his head
- Rattle of the stall dividers
- Confined space

To prepare your horse for some of these sensations, lead him across ditches and bridges, into low buildings, and under overhangs. Be sure anything you attempt to walk your horse across is designed to support a horse's weight and has no dangerous protrusions or snares. Your goal is to be able to walk your horse forward through these obstacles without hesitation and to be able to stop your horse anywhere along the way.

I've presented a group of obstacles to give you ideas for planning your own preloading obstacle training program.

In addition to the progression in this chapter, add your own obstacles. Lead your horse across concrete pads and rubber mats and tie your horse in areas that have rubber-mat flooring. Lead him into a dark building from bright light to prepare him for the sensation of entering a darker trailer, and use the exercise to note the ability of his eyes to adapt to the change in light in comparision to yours.

Expand on the backing exercises by backing your horse out of stocks, out of his stall door, out of the grooming area, and through trail obstacles.

One important aspect of trailer training is teaching your horse to stand patiently without pawing or pulling while tied to the inside or outside of the trailer. Make sure you give your horse plenty of tying experience before you haul him to his first show or trail ride. (See chapter 8 for tying inside the trailer and chapter 13 for tying to the trailer.)

A horse must accept being tied for at least 3 or 4 hours without pawing or getting nervous. Tie him out of the hot sun or a cold wind for increasingly longer periods until patience becomes habit. Although you might think that a horse would be more likely to stand still after a hard workout, a sweaty, hot horse might rub and be uncomfortable standing still.

Also, be sure to accustom the horse to all the protective gear he will be wearing for trailering — especially leg boots, tail wrap, and sheets. Do this as a separate lesson before trailering.

Backing through Barrels

7.1 Lead Alongside Barrels

Place a line of barrels about 2 feet (0.6 m) in from a safe fence. I'm using six plastic barrels to create an alley about 20 feet (6 m) long. Lead your horse into the alley by staying on the outside of the line of barrels yourself.

7.2 Back Horse Out

When you get to the end, stop the horse and stand in front of him to back him out of the alley. Take your time and let him find his way out. You want your horse to develop a habit of backing *slowly* out of the trailer, not running backward, so begin the habit here. Practice stopping halfway out of the alley. When the horse has learned to back slowly and calmly out of the alley, add clunking sounds on one of the subsequent trips, by kicking a barrel with your foot or runnng a stick or whip along the barrels as your horse backs. This will get him used to unusual noises that will occur inside the trailer. If it makes you nervous to be in front of your horse during this exercise, then hold the lead rope over the top of the barrels and stay on the opposite side. Eventually, however, you'll need to become comfortable handling your horse in a confined space. There will be times you will have to enter the trailer to attend to your horse.

Turn on the Center in a Box

7.3 Swapping Ends

Make an 8-foot-square (2-m) box out of railroad ties, as shown, or heavy poles. You might have to adjust up or down on the size of the box according to the size of your horse or pony. Mr. Mellow, a 16-hand gelding, is in a 7-foot (2-m) box. Walk your horse into the center of the box. You will begin asking for alternate steps of turn on the forehand and turn on the hindquarters. What this results in is a turn on the center, or "swapping ends," as the horse's hindquarters move to the right and the forehand moves to the left, swiveling on an imaginary point in the center of his body. Your goal is to keep the horse calm and controlled in his movements so he doesn't step out of or knock over the box.

7.4 Confusion

Your aids will be alternately asking the horse to move his hindquarters away from you and his forehand toward you. The first time, your horse might be understandably confused since you've just practiced forehand and hindquarter turns with him but never this hybrid. Just take your time and gradually rotate the horse's body 360 degrees, so you can exit the box on the same side you entered. The turn on the center will come in very handy when you want to turn a horse around in a stock or slant-load trailer, as you will be able to unload him by leading him out rather than backing him out.

Wooden Bridge

To accustom a horse to the sound and sensation of walking over a hollow floor, walk him over a wooden bridge. The training "bridge" used here is made of a 4 x 8 piece of ¾-inch (1.9-cm) exterior grade CDX plywood that has been fastened with 2-inch (5.1-cm) deck screws and glued with construction adhesive to a frame of 2 x 4s laid flat. There are 2 x 4s at 2-foot (0.6-m) intervals to support the plywood. This results in a 2¼-inch (5.7-cm) step up. It has been painted (with an oil-based primer and two coats of oil-based exterior paint) in blocks of bright colors to further develop a horse's confidence in stepping on odd surfaces. If you plan to store your bridge outside, use CCA pressure-treated wood for all bridge parts.

7.5 Taking a Look
Veteran has been over this bridge and countless other bridges many times. Yet she still looks as she steps. This is desirable — a sign of a good trail horse and of an observant horse for any use.

7.6 Taking a Bite
If you give some horses an inch, they'll take a mile. I gave Mr. Mellow a long lead so he could inspect the bridge and he seized the opportunity to do a little bit of grazing. Do not let your horse digress from the lesson no matter how cute or harmless it seems. On subsequent passes, he wasn't allowed such latitude.

7.7 Honor System
Veteran can be trusted with a lot of slack. She'll walk dead center across the bridge and not think of sneaking a bite of grass. When your horse can pass this test on a 10-foot (25-m) line, he has learned the lesson thoroughly.

7.8 Resistance

On her first approach to the bridge, Ms. Antsy Pants started to swing her hindquarters into me in an attempt to avoid going over the bridge's center. I stopped her and used a turn on the forehand to move her hindquarters back onto the track. I followed with a "Head down" review.

7.9 Compliance

Now she is walking dead center, flat footed and head down, looking but not fooling around. This is good. A horse is more confident when she knows exactly what she can and cannot do and that you are in charge.

7.10 Off-Side Cooperation

When trailering your horse, there will be times when you will have to handle your horse from the off side. Thus, conduct all lessons from both sides. Just because a horse learned something well from the near side doesn't mean she automatically transfers the lesson to the other side of her body. Often, *you* will be much less coordinated, and your horse — not being accustomed to you working on that side — might pick up on your awkwardness and react with avoidance or resistance. Ms. Antsy Pants, after being corrected going the other way, is in very good form from the off side. In anticipation of her possibly coming unglued, I have a shorter-than-normal hold on the lead rope, which doesn't allow her to put her head down. In the early lessons, it's a compromise between retaining control and giving the horse some slack, so to speak.

7.11 Excellent Form and Energy

When I give her a few inches more lead as we exit the bridge, Ms. Antsy Pants immediately lowers her head and reaches forward with her entire body in an energetic, flat-footed walk. This is excellent form and a highly successful pass for the first time leading from the off side. Should we quit here? Maybe for the time being, giving the horse a nice rewarding scratch. It is always good to end on a positive note. But plan to take the horse over the bridge many times in the coming days, with the bridge in many locations, and over other platforms before you call it a "done deal."

Platform

The platform will simulate the trailer floor in its size, step-up height, and its rubber-matted surface. This step-up platform was made by laying five railroad ties (10'6" x 8" x 6" [3.2 m x 20.3 cm x 15.2 cm]) side by side and covering them with rubber mats. This created a 10 foot 6 inch (3.2 m) long, 40 inch (1 m) wide, 6¾ inch (17 cm) high platform.

7.12 Step Up
Reflex Queen moves forward with great energy and a good rounded topline. She steps up on the center of the platform and keeps her body straight.

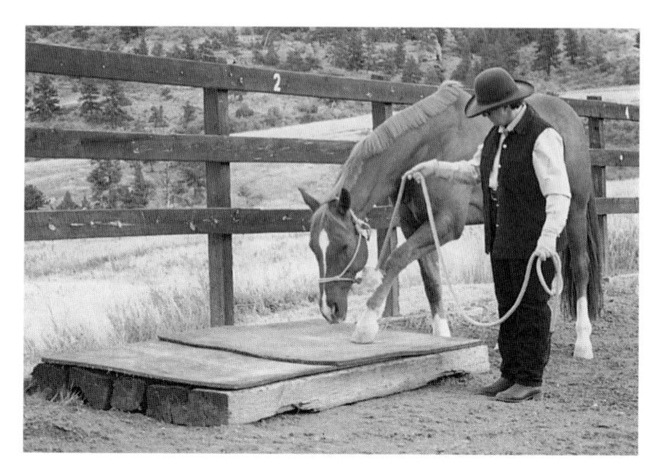

7.13 Close Inspection
Mr. Mellow, the "inspector," because of his relaxed attitude and back, can practically touch his nose to the platform as he steps up. I give him the benefit of the doubt with a slack lead despite the fact that he proved to be a grass-seeking missile on the last obstacle.

7.14 Stand
After you have led the horse up on the platform, across it and down the other side, you want to begin adding variations. The first is to stop the horse dead center on top of the platform. Ms. Antsy Pants is standing square and confidently on a slack line. Prior work on developing patience is paying off.

Once a horse walks onto the center of the platform and will stand on a slack line, it's time to begin the backing-off lesson. Start from the near side first.

Backing Off the Platform from the Off Side

7.15 Her Customary Mild Resistance

The first time Ms. Antsy Pants was asked to back off the platform from the off side, she raised her head, stiffened her neck, and tilted her head. She threw the weight of her forehand to the left and moved her hindquarters to the right, just about to step off the side of the platform. *Stop.* This is not productive.

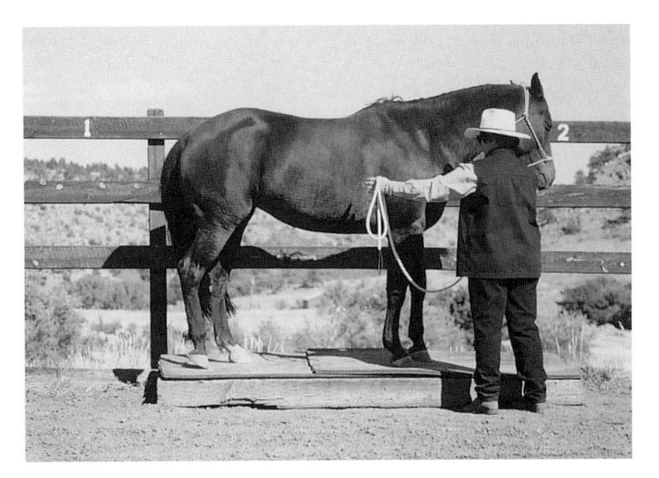

7.16 Relax, Give Guidelines

I lead her back up onto the platform, but before I begin backing her again, I lower her head using slight pressure on the poll and noseband of the halter with my right hand on the lead rope. This causes her topline to round, which will put her in better shape to back off the platform. I extend my left arm to give her visual guidance and a "warning" not to step to her right but to back straight off the platform.

7.17 Good Form

We maintain our positions, and at the midpoint she is still in the center of the platform and has a lowered head and neck. Ms. Antsy Pants has maintained her straightness and rounded topline.

7.18 Nipped in the Bud

Now she is ready to take the last steps off the platform. This is a perfect example of how a correction is a necessary part of learning. This filly has been demonstrating in this and other lessons that after one correction, she performs well. Had she been allowed to back off crooked the first time, it might have been more difficult to make the mental and physical connection with her later to change the behavior. Try to nip little mistakes in the bud.

Adding Trailer Boots

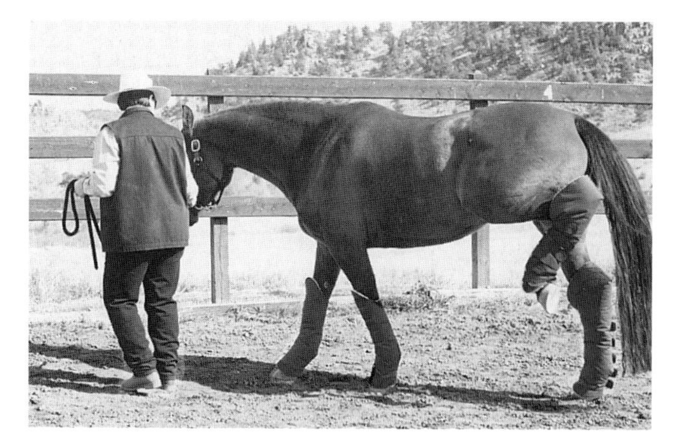

7.19 EXAGGERATED FLEXION
Some horses have a strong bucking, kicking, and striking reaction to anything touching their legs. The Reflex Queen is a sweet-tempered filly, but she has very exaggerated reflexes. I lead her with great attention to my safety. Note how she hyperflexes her hind leg in reaction to the boots, even though this is the second lesson with them. Her reaction the first time was so violent that I thought she was going to injure herself or remodel the facilities.

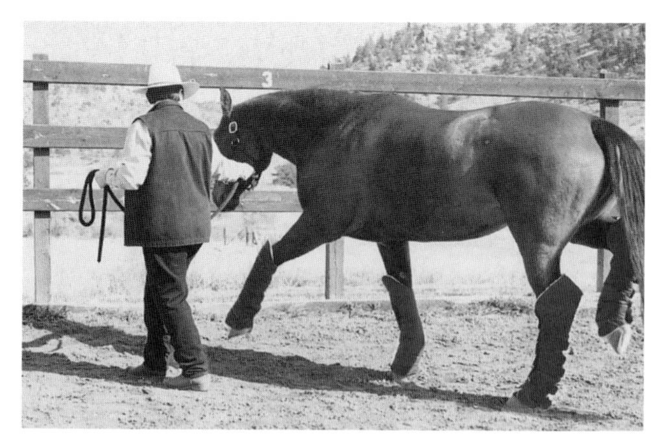

7.20 GOOSE-STEPPING
Here, even at a walk, the flexion of her movement is very accentuated. She is becoming accustomed to the sensation of the boots rubbing her hair and creating new pressures as she moves. Maintain your position at the side of the horse and be careful that the horse doesn't accidentally strike you.

By the time we get to the platform, her reactions have toned down somewhat, but Reflex Queen is still goose-stepping considerably.

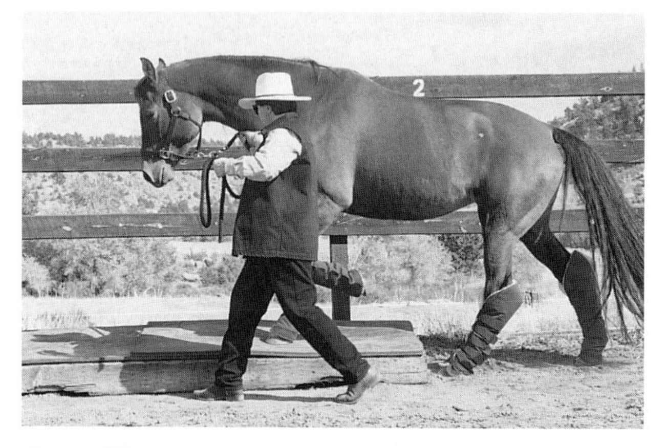

7.21 FOCUSING ON THE PLATFORM
There is no hesitation when asked to step onto the platform because she had been taken over it a number of times without boots. This is a good example of how a progressive method of getting one lesson down pat (walking over the platform) will make subsequent lessons (walking over a platform while wearing boots) go more smoothly.

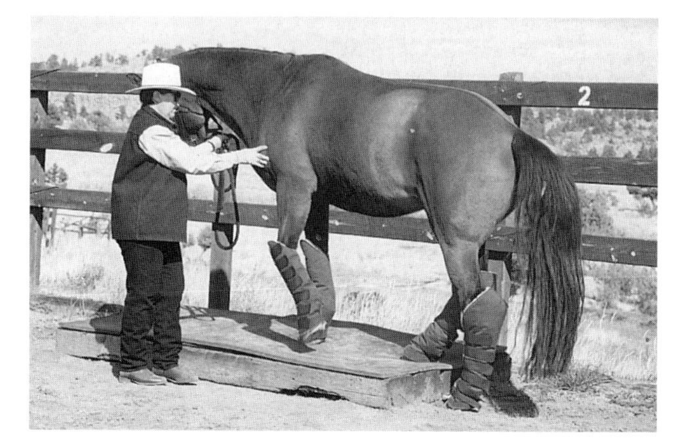

7.22 BACKING WITH BOOTS
The sensations caused by backing in boots are different from those in a forward movement and can cause a horse to kick out, so be sure to practice backing thoroughly as well. I am using the shoulder cue here.

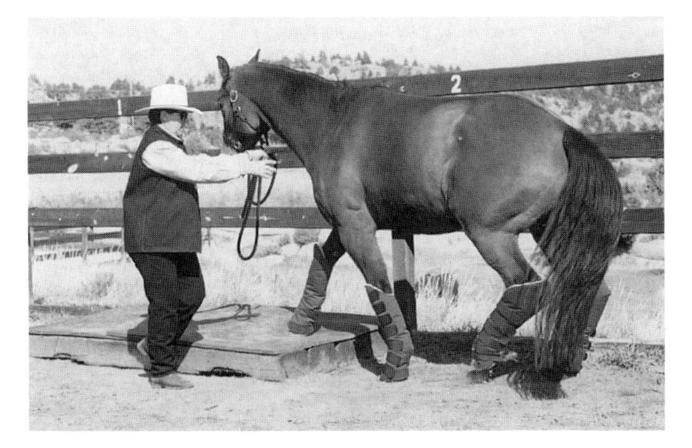

7.23 Conditioned to Boots

The back is completed with a visual cue from my right hand but no touch. Reflex Queen remains straight and she has forgotten about those odd things on her legs. The boot lesson must be repeated a number of times for this horse to become thoroughly conditioned to wearing them.

7.24 Step Down, Head Up

This photo demonstrates how high Mr. Mellow's head comes up when he takes that first step off the platform or a trailer. Picture this horse in a trailer. Well, it had better be a tall enough trailer or he'll be hitting his head on the ceiling. It takes only one instance of a horse hitting his head to make him shy about going into or coming out of a trailer.

Leading Under a Tarp

7.25 Covered Alley

Lead your horse under a safe low ceiling or a temporary tarp to accustom him to something directly over his head. To make this temporary roof, I've put a few metal panels in my arena and draped a tarp between the metal panels and the arena fence. The roof is about 5½ feet (1.7 m) high at the low end and 7 feet high at the panel end. When walking 15-hand Veteran under the tarp, there is plenty of clearance and no concern. If a horse did raise her head, the soft roof wouldn't injure her.

7.26 Two Ducks

Here's another view of the same setup and horse. I have to duck a little to fit under the low end and Veteran ducks as well, even though she has plenty of clearance. The alleyway is about 5 feet (1.5 m) wide. I give the horse a good amount of slack in the rope to allow her to lower her head. Once a horse walks calmly through, stop in the middle. Then, have someone rattle the panels as you pass to simulate trailer divider noises.

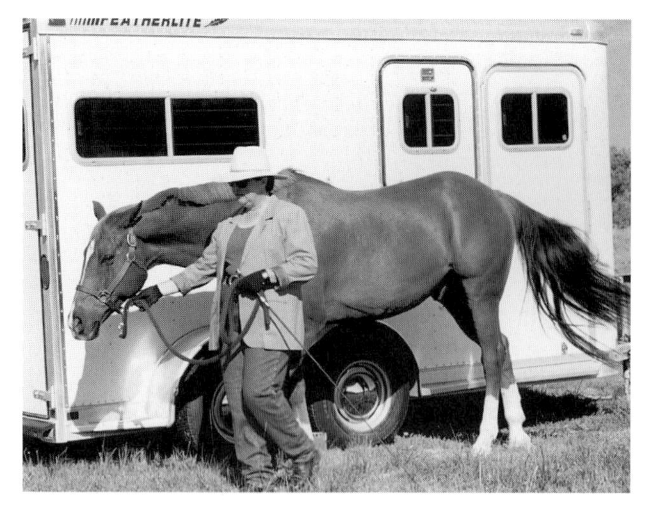

7.27 No Big Deal

As part of the in-hand obstacle training, lead your horse past all four sides of the trailer, leading from both the near and the off sides. Mr. Mellow is unconcerned, as most horses are. Even if a horse has never been loaded, if he has seen trailers parked on the property, generally they pose no big threat.

7.28 Meet the Trailer

Rookie has spent the majority of her life on pasture and has just been brought in for a bit of handling at 12 months of age. I've fitted her with a rope halter. When first asked to approach the trailer, she stops dead in her tracks. She is not interested in looking at or sniffing the object. After waiting a few moments, I initiate sideways and forward movement by popping her with the end of the lead. I want her to move her hindquarters toward the trailer and to step forward at the same time. This photo captures the moment when I have started my step forward and the lead rope has yet to land on her side.

7.29 Spooky Shadow

Here is Rookie's reaction. She does move her hindquarters sideways and she does move forward. However, when she "arrives," she appears terrified at what she sees out of her right eye. I'm using my body to block her from pushing me over and am taking great care not to get stepped on. She is either frightened of her shadow on the trailer or the glaring white surface in the morning sun.

7.30 CHECK IT OUT
Horses are basically curious and want to investigate things. I give Rookie a second chance to sniff and touch the trailer to allay her fears. She is still on "ready alert" here, as evidenced by the fact that her right hind leg isn't bearing any weight. I always give a horse the benefit of the doubt and allow slack in the lead rope, which enables her to move naturally and to extend her neck.

7.31 "GOOD GIRL"
When we step up to the trailer window, note that Rookie has weight on all four feet, her eye is softening, and she is still inspecting the trailer. As I reach with my right hand to stroke her neck, I give her even more slack in the lead rope so she feels free to reach for the window hinges. Unbeknownst to either of us, Smitmo the cat has parked himself beneath the edge of the trailer right at our feet.

7.32 STEP CLOSER
Now that the issue of the trailer has been settled, I review the turn on the forehand lesson. I am asking her to take just one step to the right so that her hindquarters are closer to the trailer. I do not hold Rookie's head forcibly in position, but still give her slack in the lead rope as I move her hindquarters over with a flick of the lead rope. She's not 100 percent calm yet about all of this, but I don't feel she's going to bolt again. Rookie seems unconcerned about the presence of Smitmo the cat, even though he is in one of her blind spots, below and slightly in front of her head. However, I send the cat away, for everyone's safety, before we continue.

7.33 Walk Parallel to Trailer
Once I get her body completely parallel to the trailer, I walk Rookie past the trailer. I'm holding the lead rope about 10 inches (25 cm) from the halter, which is a compromise between control and the honor system. Her neck is still a bit tense on this first pass.

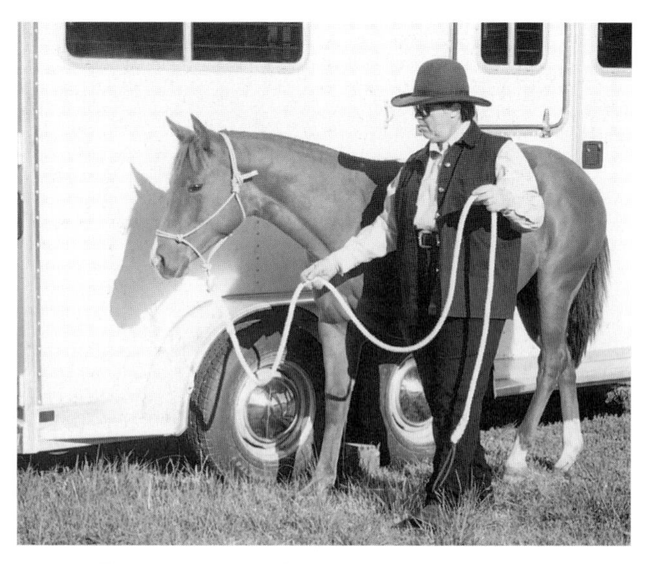

7.34 Plenty of Slack
When I read that Rookie is ready for the final test, I give her 2 feet (0.6 m) of slack in the lead rope and she stays in a very good position. She relaxes her neck and lowers her head to the perfect level that allows her to see what is going on and lets her use her back and hindquarters in relaxed strides.

7.35 Web Halter and Chain Option
This entire progression could have been conducted with a halter and chain (with the chain either parked or engaged) and an in-hand whip. The whip allows you to reach farther back on the horse's hindquarters, and you can deliver the cue more subtly than you do when you pop the horse with the end of a lead rope. Choose what feels comfortable for you and what works consistently. The principles are the most important aspect of any training, not the tack.

Leading and Backing in Trailer Alley

7.36 Stop in an Alley
For the final preloading obstacle, park your trailer alongside a building or safe fence and create a 5-foot-wide (1.5-m) alley. Walk your horse through this alley from both directions. When your horse walks through without quickening her rhythm, stop in the middle of the alley. Then walk out calmly.

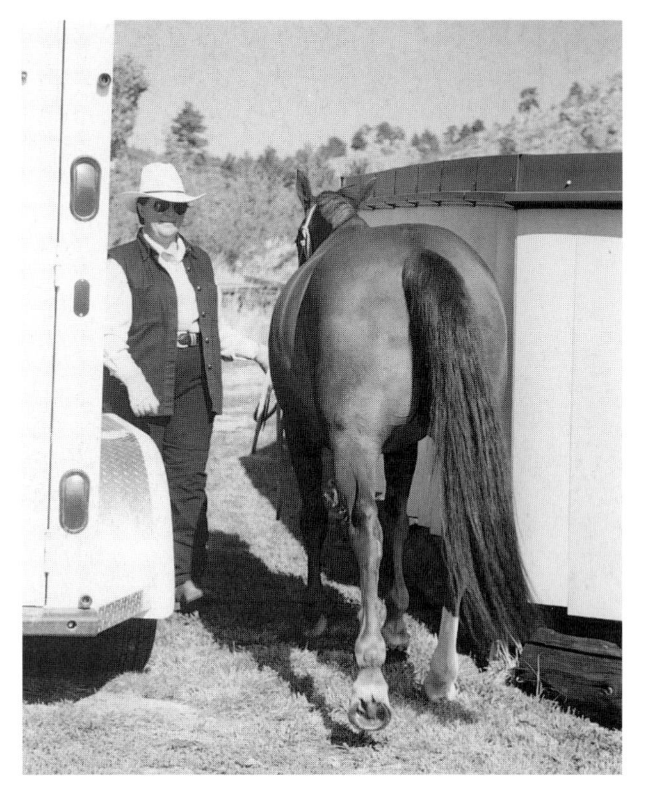

7.37 Back Out of the Alley
Finally, stop your horse in the alley and back her out. By this time, you should not have to use any physical cues to back your horse: Just the command "Baaack" should suffice. Be sure the horse backs straight and calmly. Do not let the horse rush out of the space backward.

8

✦ TRAILER LOADING AND UNLOADING ✦

If you have taken your horse through all of the lessons in chapters 6 and 7, leading or sending him into a trailer should be a piece of cake. Before you attempt to load a horse that you are training, take a seasoned horse into the trailer so you become familiar with the trailer space, its rattles, and its design.

With a straight-load trailer, the safest way to train a horse to load and unload is to send him in and out. For many slant-loads, stock trailers, four- and six-horse trailers, and vans, it doesn't work really well to send a horse into the trailer. Therefore, I've also included instructions on leading in and leading out. Whenever you enter a trailer alongside a horse or a trailer that already contains a horse, you run the risk of getting hurt if the horse panics. Choose the methods that work the best with your experience, the horse's level of training, his temperament, the style of your trailer or van, and the size of your horse in relation to the trailer.

All of my methods are low in stress, progressive, and involve minimal cuing. Because I work alone most of the time, I've designed the methods for one person, either male or female. You don't need super strength or specialized tack to have success in loading a horse.

Remember, encourage your horse to take one step at a time. Be patient and take the time to make your horse a deliberate, safe loader and unloader. Prevent dangerous bolting in or out of the trailer — both are much more difficult to cure than to prevent.

Before you load a horse into a trailer, use the following checklist to make the trailer safe and more inviting for the horse. (More details will be given in the trailer-loading sequences.) Tailor this list according to the style of your trailer and your horse's experience and needs.

1. Make sure that the trailer is fastened to a towing vehicle or a farm tractor for practice, otherwise the weight of the horse stepping on the back of the trailer will likely cause the front of the trailer to come off the ground. *Never* load a horse into a trailer that is not connected to a vehicle.

2. Open the manger doors or drop-down head doors and fasten them in the open position. This will allow more light and air into the trailer so it won't appear as confining.

3. Open the rear door as far as it will go and fasten it securely in this position. If there are two rear doors, open them as far as they will go and then fasten them securely. A door slamming on your horse from a wind gust just as he is about to step into the trailer is guaranteed to make him leery of the trailer. (See chapter 3 for more information on door fasteners.)

4. In a slant-load, fasten all dividers securely against the wall so they don't swing toward the horse just as he steps in.

5. In a straight-load, consider moving or removing the center divider to give the horse more room. This might be necessary temporarily during training or permanently for a large horse.

6. Place a railroad tie under the sill of the trailer. This will be especially useful when unloading to prevent a hind leg from slipping under the sill.

7. Place a treat in the manger or feed bag of the trailer so the horse will instantly be rewarded for loading. Never use treats or feed to bribe a horse into the trailer. Just let the horse discover the treat after he has stepped fully into the trailer.

Loading in an Open Trailer

When Mr. Mellow was 2 months old, a friend loaned me her trailer so I could demonstrate training a foal to load using a ramp (photo 8.1). The next time he was loaded in a trailer was in the photos for this book! Time flies and horses grow. By the time he was a yearling, he was too tall for my very short, 1974 stock-horse trailer. By the time I upgraded to a taller trailer, this tiny foal had grown to 16 hands! His first trailer-loading experiences since photo 8.1 are chronicled beginning with photo 8.2. Even though he hasn't been loaded, he has been longed and ridden and has been a photo model in my other training and care books.

Photo 8.1 shows good and bad things. The good thing is that because the foal had extensive in-hand work, he loaded readily, even with his mother watching from outside the trailer. The potential dangers are related to the narrow, steep ramp, and the narrow passageway that the handler and horse must pass through to enter this trailer. Although there are a number of trailers configured this way and they are fine when things go right, when things go wrong, you can have major problems. The horse could slip off the ramp and possibly skin his legs, and the handler could be smashed if the horse bolts as he enters the narrow passageway. There are much better trailer choices today, as you saw in chapters 2 and 3.

8.1 Foal Demo

I load 2-month-old Mr. Mellow while his dam watches.

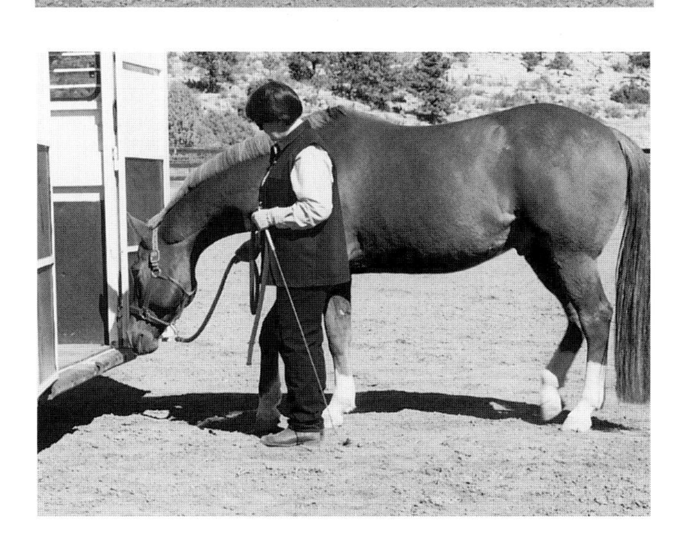

8.2 Touch and Sniff

With the trailer prepared as outlined at the beginning of this chapter, we approach. Mr. Mellow is a 16-hand horse and the trailer step up is 12½ inches (32 cm). The trailer has a rubber bumper guard on the sill. This is a slant-load trailer with the dividers fastened along the wall, so the interior is just one large open space. I give the horse 18 inches (46 cm) of slack on the lead rope. He lowers his head to look inside as we approach. I hold my whip in a lowered position. I will not urge the horse forward at this time if he wants to stop and inspect.

As is his style, Mr. Mellow likes to touch and sniff, so I allow him to gauge the height of the step and to inspect the floor.

Loading in an Open Trailer (continued)

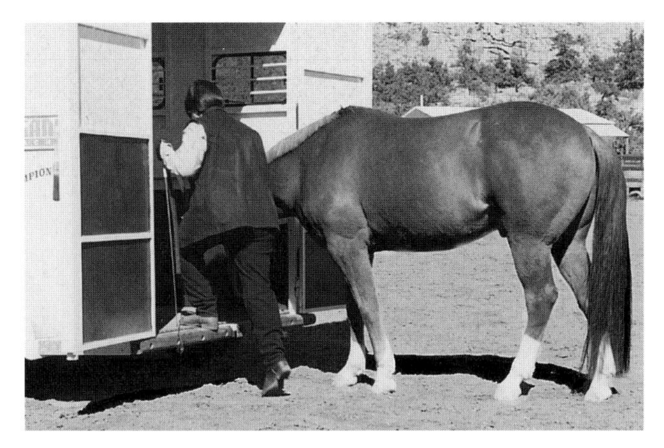

8.3 Step In

I step into the trailer.

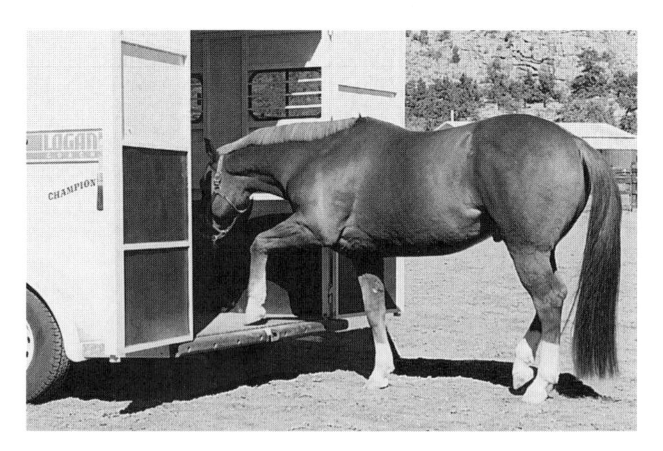

8.4 Walk On

I use the voice command, "Walk on," and Mr. Mellow follows me inside.

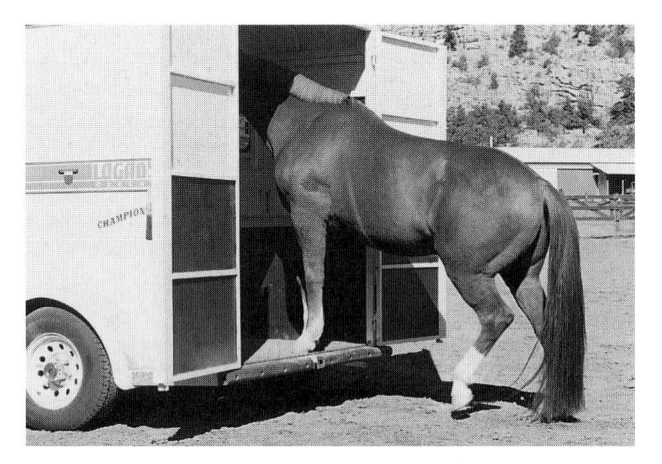

8.5 One Step at a Time

It is all right if your horse stops with just his front feet in to look some more. Even if he backs out, it's okay because you want to take things slowly and one step at a time. This horse didn't stop; he just kept coming in, as you can see by his left hind in motion. Later you will want to intentionally stop your horse halfway in and then halfway out when backing out. This will boost your confidence as a handler and reinforce to the horse that you are in control.

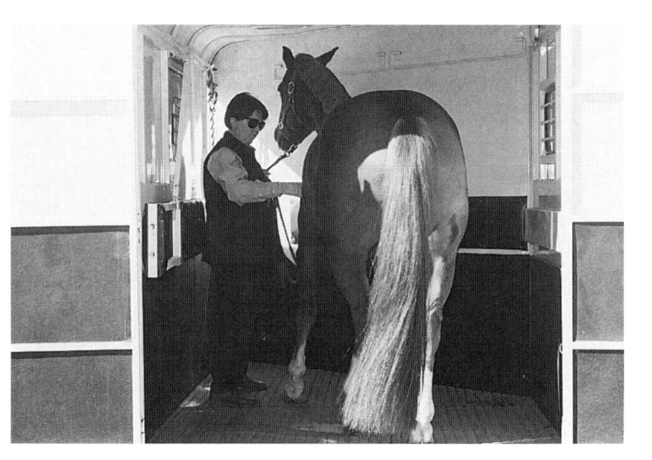

8.6 Positioning for the Turn

Once inside, I check his position to be sure he has room to begin the turn-around. To give him space for his head and neck to turn left, I first need to move his body away from me one step. With slight pressure on the lead rope to tip his nose to the left, I alternately exert pressure on his left shoulder and rib cage with my right hand.

Next we begin a turn on the center. With continual positioning of his head and neck to the left, I move my hand to a mid-rib position and ask for alternate steps to the left with his front legs and to the right with his hind legs.

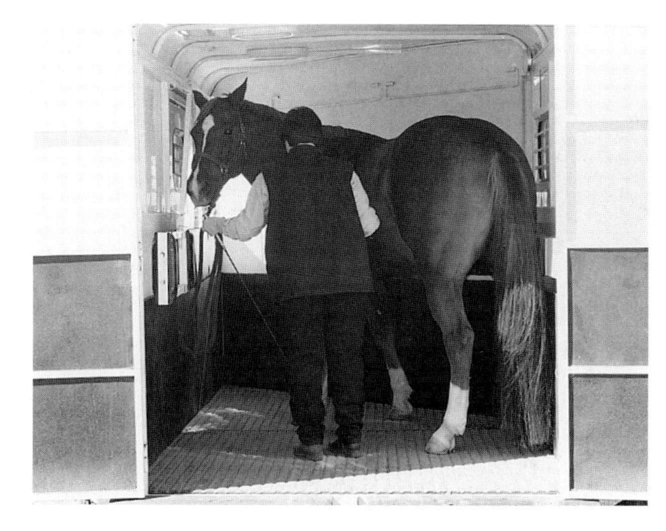

8.7 The Tight Spot

We are approaching the point of no return. Here's where a horse can feel cramped and become concerned. That's why the work in the box, on the pad, and under the tarp is so helpful. If a horse is going to blow, it will be about now. Besides asking the horse to turn sharply to the left and respond to my sideways cues, I'm also going to ask for a step forward. I want his legs to step toward the window while his head is curling around toward me. This would be very difficult for a horse to understand if he had not had in-hand work.

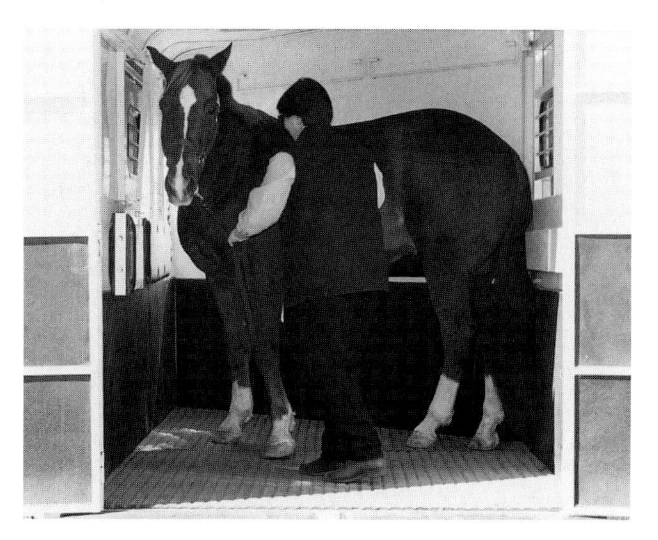

8.8 Finishing Up

Here you can see how this horse has arced his body to the left, conforming to the confining walls of the trailer. Now all that is left is to ask him to walk forward at the same time that he is performing a turn on the forehand.

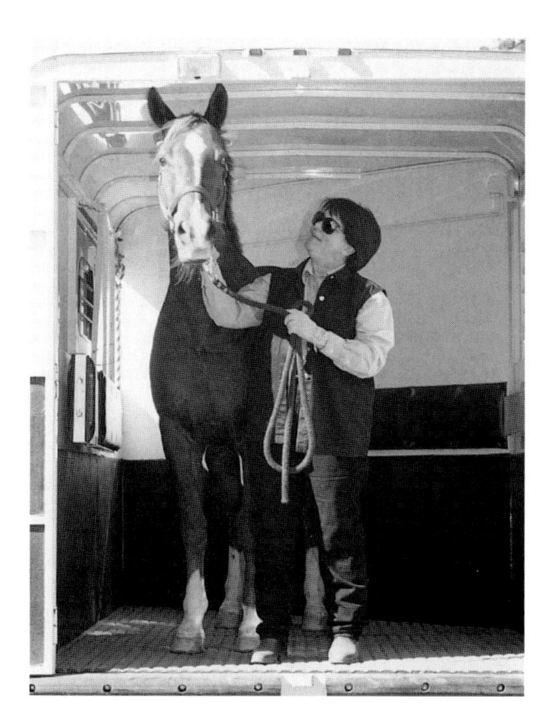

8.9 Pause and Relax

When we are finished turning around, Mr. Mellow demonstrates how low he was holding his head for the entire procedure. When he stands naturally, his ears touch the ceiling of the trailer. The success of this venture began with the head-down lesson and involved each and every in-hand and obstacle described in the previous chapters. You can be equally successful with your horse if you follow a progression and build on it. Once your horse has turned around completely, pause for a considerable amount of time. You want to be sure that your horse never gets the idea that he's allowed to rush out of a trailer. Be sure it is you who tells him when he can step out. (See "Unloading" later in this chapter and "Avoiding Bad Habits" on pages 110–112.)

Finishing Up with the Slant-Load

When it comes time to travel, you'll need to tie your horse and hold him in position with a stall divider. (If you are hauling just one horse, you can leave the divider fastened to the wall and let the horse have the whole open space.) Before you lead your horse into the trailer, make sure you have a tie rope fastened to the tie ring near the ceiling. (See "Tying," later in this chapter, for complete details on tying in a trailer.) After you've led your horse into the trailer, hand the lead rope to a helper who is standing outside the trailer window. Then fasten the divider and the rear doors. Finally, snap the horse's halter to the trailer tie and remove his lead rope.

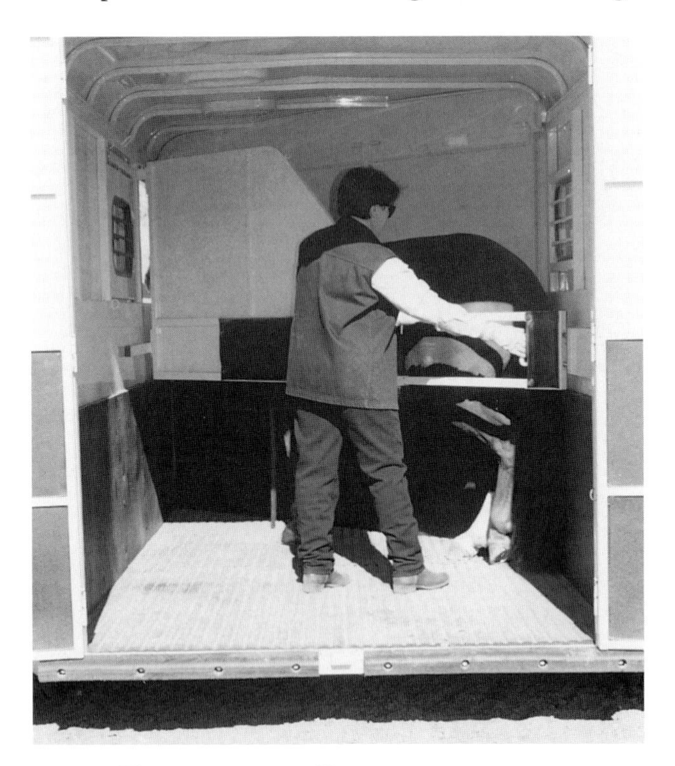

8.10 Fasten the Divider
Unfasten the stall divider from the wall and swing in toward your horse, taking care not to bump or startle him. Fasten the divider into its slot in the wall. Note that this divider has a solid panel between the head portions of the two stalls so that the horses don't fight or play in transit. Once you have fastened the stall divider, close the rear door(s), then tie your horse.

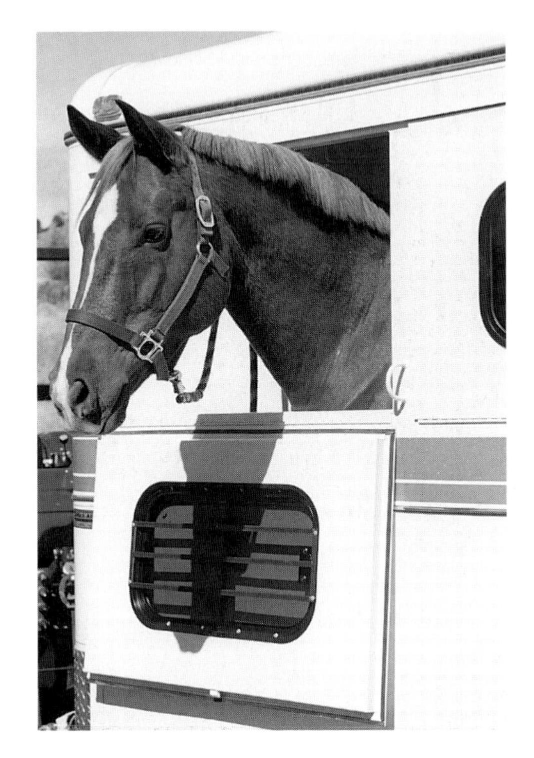

8.11 Head Out 'til It's Time to Go
By leaving the window open, you allow the horse to put his head out, which gives him something to keep him occupied while you are fastening the stall divider. It also gives him a lot more room lengthwise. When you are ready to travel, he will have to back up until his head is inside the trailer. Be sure your trailer is long enough for your horse. *Never* let the horse put his head out the window when you are traveling, however. It will almost certainly result in eye injury from airborne debris; it could also end in death from impact with any number of items along the road.

Unloading from an Open Trailer

When you unload a horse from a slant-load or a stock trailer, either turn the horse around and lead him out or back him out. (Refer to the previous photos on turning the horse around.) Once he is facing out the back of the trailer and his body is straight, pause before you step down or ask him to step down. Then let him have enough slack in the lead rope that he can put his head down to balance and to look where he will be stepping.

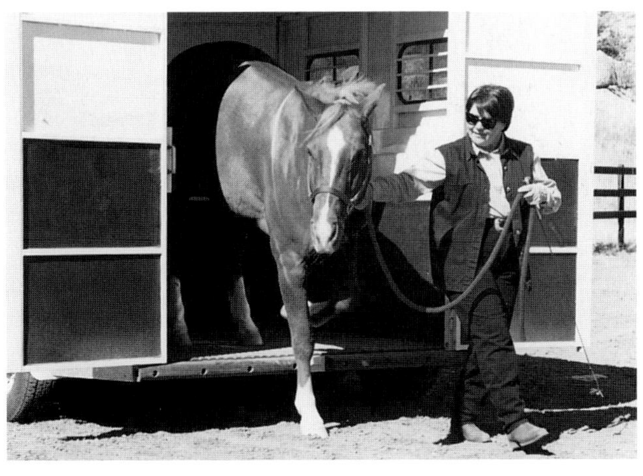

8.12 Lead Out

Let him step down slowly and carefully. There is no hurry. Allow him to dictate the pace, as long as it is slow and measured. Keep your eye on the horse when unloading because horses will become aware of the new scene that suddenly appears to the side and may not pay close attention to your position.

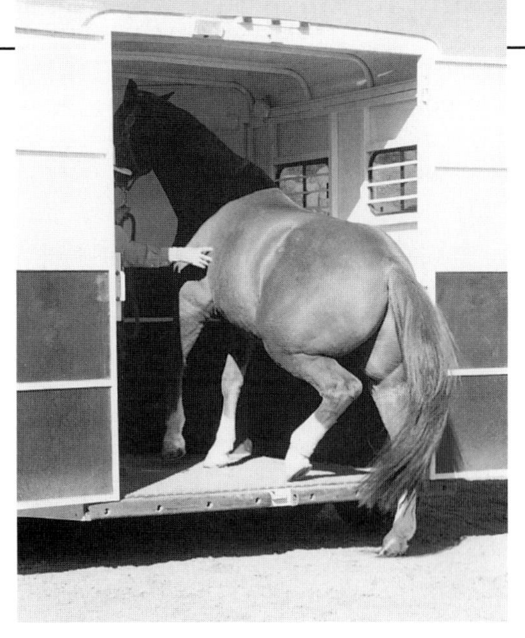

8.13 Back Out

If a horse has previously been turned around and walked out of a stock trailer or slant-load, when you try to back him out he may try to turn as he is unloading. Be aware of this and keep him straight by using your hand alongside his ribs to indicate that he is not to move toward you as he backs. Let him take his time. If he stops halfway, that's okay. If he walks back in after stepping down, let him. He needs to feel out the situation. Many horses, when prepared with in-hand work and obstacles, just back out without hesitation. Note that when this 16-hand horse steps down with his hind leg, his ears touch the ceiling. I'd suggest a taller trailer for a horse of this height.

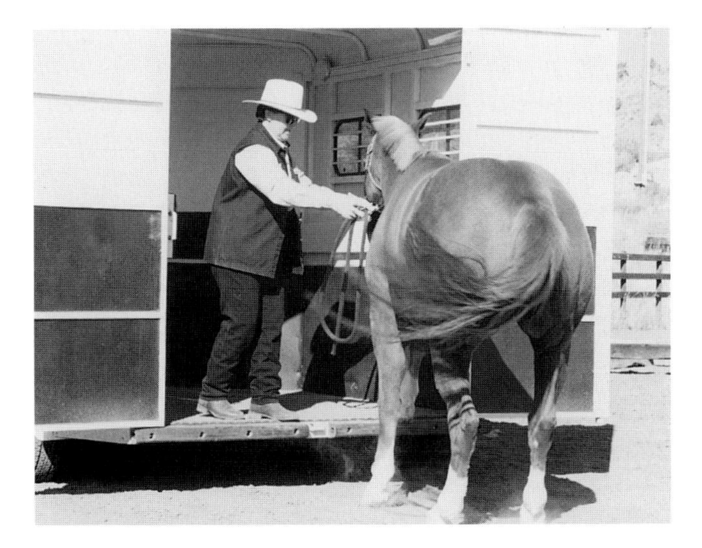

8.14 One Step at a Time

Mr. Mellow has three feet on the ground, his body has lowered, but his last hoof is "stuck" and slowly dragging off the floor. Once a horse gets three legs on the ground, it is difficult for him to pick the remaining foot up.

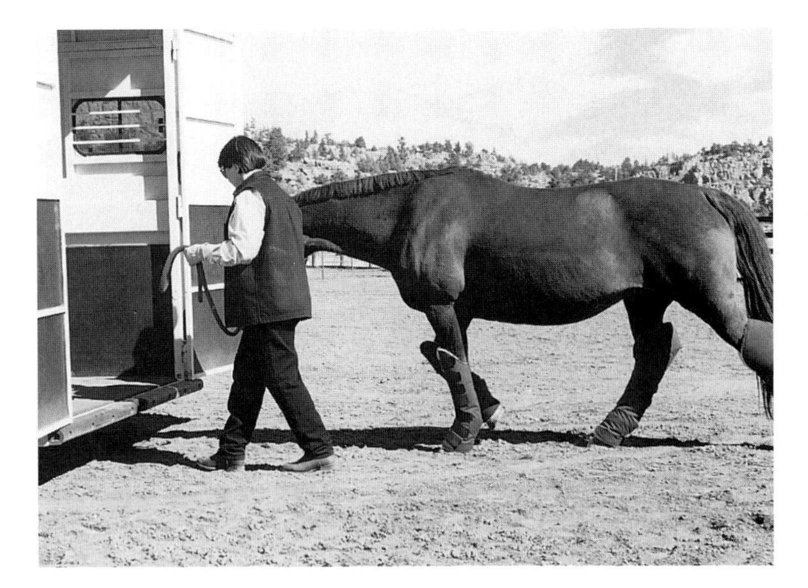

8.15 Practice with Boots
Take the time to outfit your horse in boots and lead her into the trailer as part of training so that on the day of your first trip, you won't need a crash course in traveling in boots.

Trailering Tip

Don't forget that your training isn't complete until you outfit your horse in trailer boots (see chapter 11 for more information). Trailer boots are much more convenient than leg wraps, and many styles protect the horse's knees and hocks from bangs.

Loading in a Straight-Load — Demonstration of Goal

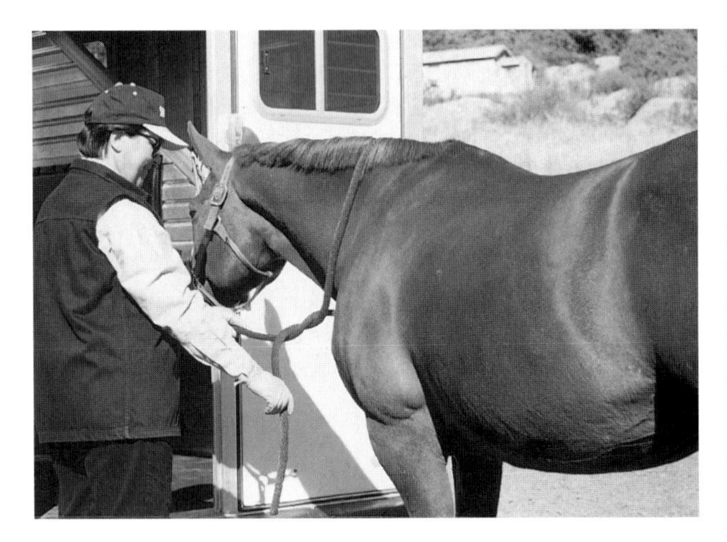

8.16 Lead Rope Over Neck
The procedure for loading Veteran in a straight-load and the goal that you are heading for with your horse in training are outlined in the next set of photos. With the trailer prepared as described at the beginning of this chapter, take the horse up to within a few feet of the trailer opening. Be sure the horse's body is straight when she faces the opening. In this case, because Veteran is a seasoned loader, the stall divider is in its normal place at the center and the right door is closed. Tell the horse "Whoa" as you drape the lead rope over her neck and then take a wrap or two around the rope. This will prevent it from falling off the horse's neck just as she is about to step up. If a horse steps on her lead rope, it can cause unnecessary confusion, as the horse would be receiving conflicting signals — go and whoa.

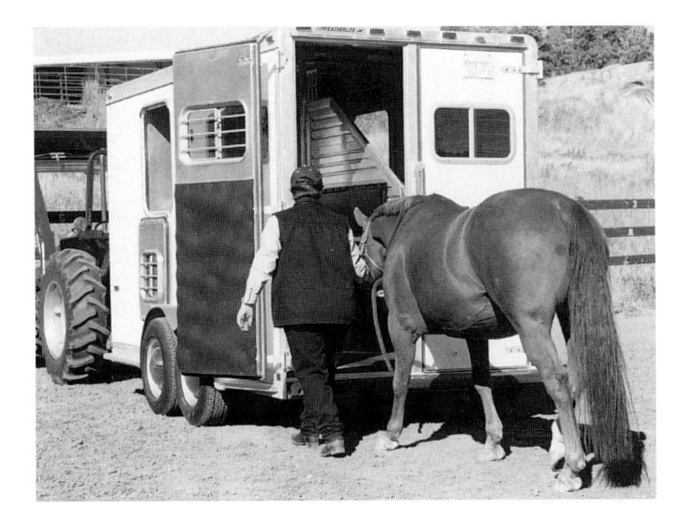

8.17 "WALK ON"

Now, taking the halter sidepiece in your hand, say "Walk on" and send the horse into the trailer. With some horses, it helps if you take one step yourself to get them started. A horse that really knows "Whoa" will often want a definite cue from you that it is now time to move forward. It's like a formal go-ahead.

8.18 RELEASE HORSE

Release the halter as the horse takes the first step. Stay in a safe position at the side of the horse. As the horse is taking her last step up into the trailer, you can begin reaching for the butt bar, which is located along the wall of the trailer.

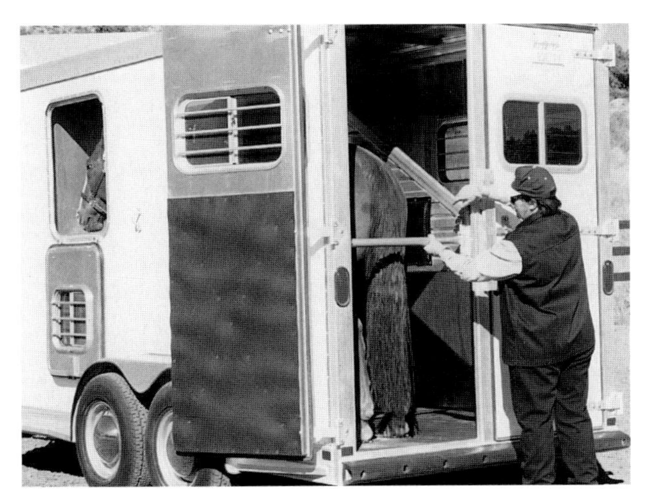

8.19 FASTEN BUTT BAR

Swing the butt bar over to the center divider and fasten it. Stand off to the side behind the other half of the trailer. If you stand directly behind your horse, and she all of a sudden bolts backward because you forgot to check for hornet's nests in the manger, for example, you could be killed. Practice hooking the butt bar without a horse in the trailer until you can do it smoothly with your eyes closed. Any problems with the butt-bar mechanism should be fixed before loading a horse.

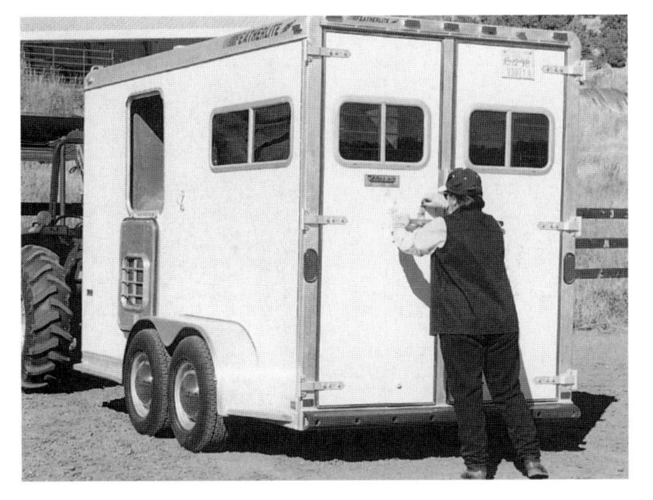

8.20 CLOSE DOORS

Close the door and fasten it securely. Be sure your trailer doors have positive latches on them. If there is a place for an additional fastener to keep the latch closed, use it. (See chapter 3 for more on latches.)

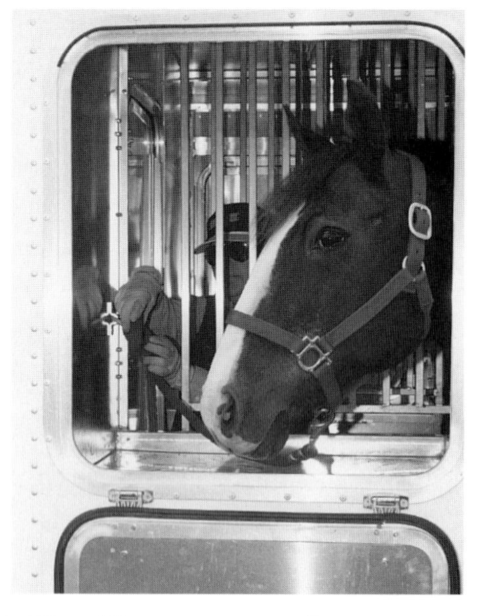

8.21 Tie Horse from Inside

Only after the butt bar is fastened and the trailer doors are securely closed should you tie your horse. If you tie a horse before the butt bar is fastened and the horse pulls back, there could be a disaster. He could break the halter and come flying out, or if the halter didn't break, he could seriously injure his hind legs on the back edge of the trailer and hit his head and front legs on the manger if he pops forward. This would create a serious problem loader. You can enter the empty stall of the trailer and tie the horse from inside. (See "Tying," pages 103–105.) Here I am using the horse's normal lead rope to tie her in the trailer. This is okay, but it has a few drawbacks. For one, the lead rope has a bull snap, not a panic snap. Therefore, if you use this method, tie with a quick-release knot on the manger ring. If you get in an accident, you want to be able to release your horse quickly. Also, as the horse eats, the lead rope gets covered with hay and slobber. Later it won't be nice to handle.

8.22 Tie Horse from Outside

You can tie your horse from the outside by reaching through the trailer window. First I'm unsnapping the lead rope from her halter and removing it from around her neck. Then I will use the trailer tie with panic snap (which is hanging out of the window) to fasten her. (See "Tying," pages 103–105.)

Leading-in Tips

In some instances you might need to lead your horse into a straight-load. This always adds some risk for you, so take care and always use an assistant. That person's role is to latch the butt bar and rear door while you hold the horse at the manger before tying. (*Note:* When leading a horse into a slant-load, an assistant can stand outside the head window to help.)

If your trailer has a low center divider, you can lead the horse into one stall by walking in the other stall yourself.

If you have an escape door, you can lead the horse into a stall by walking in ahead of him and leaving via the escape door. There are two problems involved in using an escape door, however. Anytime you walk into a trailer ahead of your horse, you risk getting trampled if the horse suddenly lunges in. And, of course, you don't have the training control that you do when you are working alongside him. Also, when using an escape door, some horses try to follow the human right out the door and can become wedged or panic-stricken.

If you have a walk-through door to the tack room, you can lead the horse into the trailer by going in the stall ahead of the horse and leaving via the tack room. Again, there is the risk of being trampled if the horse rushes in.

Loading in a Straight-Load — Training Session

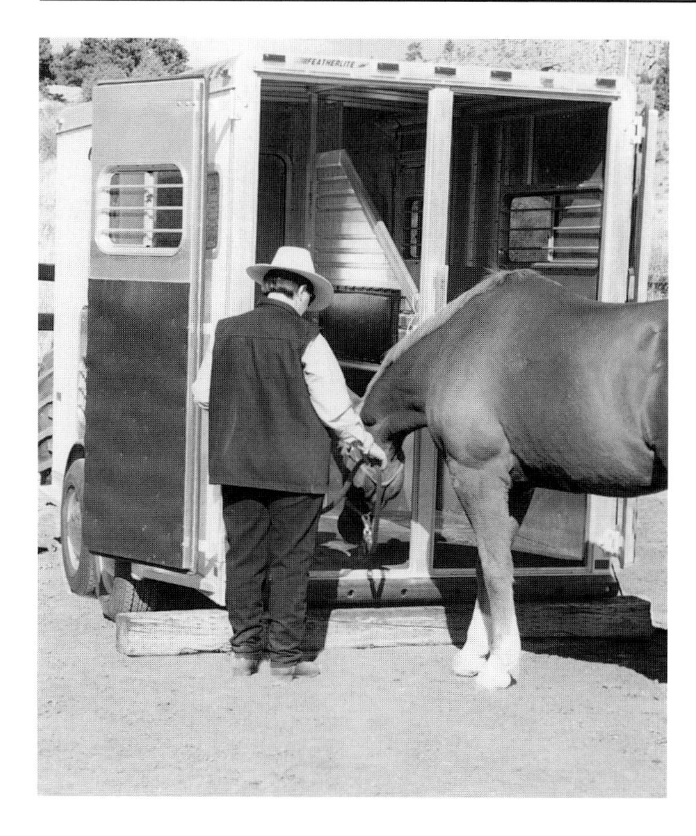

8.23 Touch and Sniff

For this training session, I am not using an assistant because I am only teaching the horse to walk into the trailer, stand quietly, and then back out of the trailer when I request. I will not be closing the rear doors or tying the horse. This lesson is a stepping-stone on the way to the main goal, which is sending the horse into the trailer.

In this first straight-load session for Mr. Mellow, he starts as with all other lessons: touch and sniff. I've opened both doors to give the trailer a greater feeling of openness. I plan to lead Mr. Mellow in and back him out until I feel he is ready to go in on his own. (This trailer has a walk-through to the tack room.)

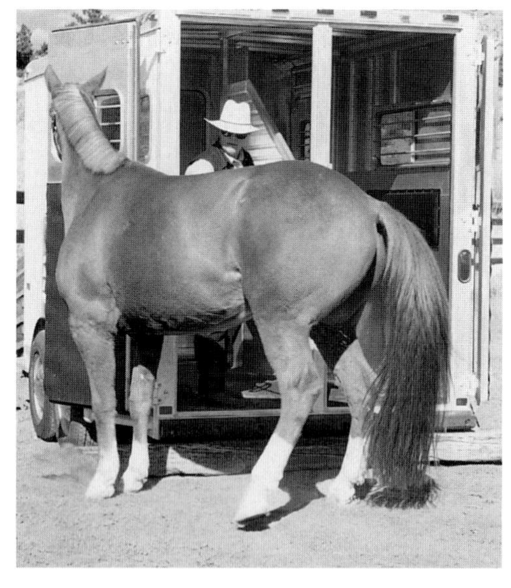

8.24 No Thanks!
When I enter the trailer, I give Mr. Mellow a long lead so there can be quite a distance between us for safety. The center divider rattles a bit, and the horse uses the slack in the lead rope to take a large step to the side.

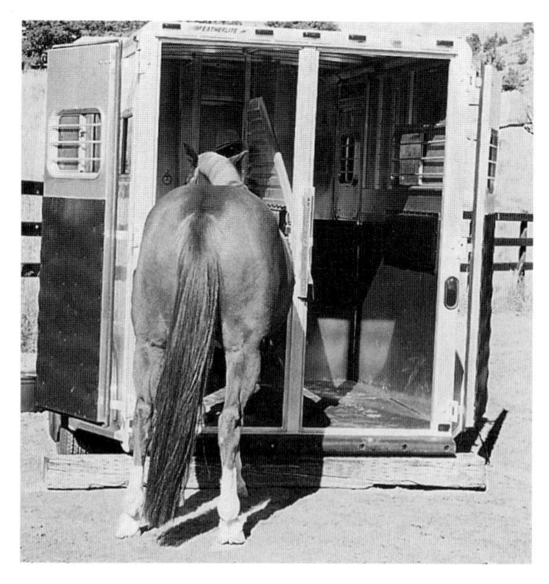

8.25 Get Centered and Try Again
I straighten the gelding, tell him "Whoa," and then I step all the way to the front of the trailer so that I am completely out of his stall. The whoa on the long line lesson pays off here. I let him stand and study the space a bit. I don't speak to him. He is on "Whoa" until I tell him otherwise.

8.26 Okay
When I'm ready, I use my most spirited and inviting "Okay. Walk on," almost chirping as I say it. Mr. Mellow loads in one continuous move. He finds the treat that I previously placed in the manger. I praise him and we stand quietly for several minutes. I do not tie him. (*Note:* If he wanted to back out, I would let him by giving him slack in the lead rope. Then after we settled and reorganized, I'd ask him to step back in.) Mr. Mellow was very content to stay inside the trailer with me. After a few minutes, I unloaded him. (See "Unloading," page 100.)

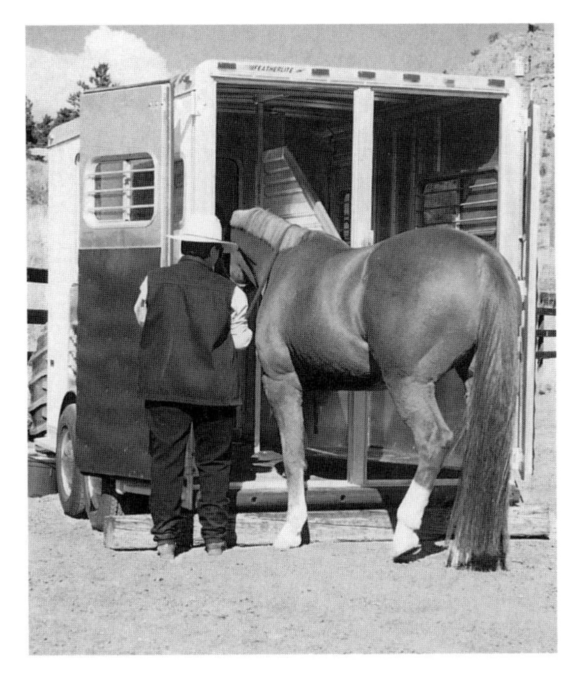

8.27 Send Him In

Because I don't want to lead my horses into a straight-load trailer unless absolutely necessary, I quickly want to teach them to load by themselves. Sometimes I skip the lead-in lesson completely — it depends on the horse and the trailer. In this case, after one lead in, I prepare to send Mr. Mellow in by himself. The lead rope is draped safely over his neck. This photo shows the first step up with his right front foot. You can see there is calm, forward motion because of the action of the left hind leg. I stand safely to the side. Avoid the temptation to continually say "walk on." Once should be sufficient as long as the horse is moving forward, albeit slowly.

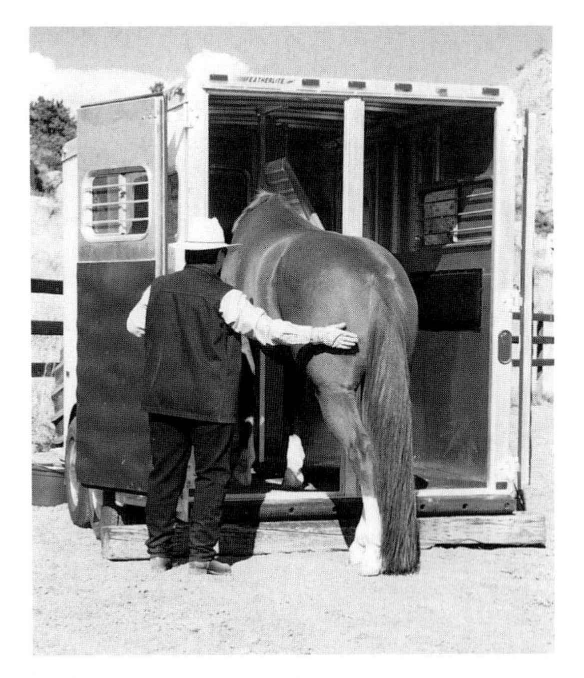

8.28 The Guiding Hand

Here is step two. Mr. Mellow is continuing to inspect. He has slowed down but hasn't lost his forward motion because he's stepped forward with his right hind. I haven't touched him yet, but I have my right hand ready to give him a pat if it looks like he might want to back up. Often a gentle reminder like this will prevent a bad habit from starting.

Note: As I have mentioned in previous sections, if a horse wants to back out of a trailer when he is learning to load, that's fine. I let him. But if I sense that the horse can be persuaded not to back merely by the presence of my hand, I'd rather give him that little reminder than have him back out. It is not necessary to use whips, ropes, or scare tactics to make a horse enter a trailer.

8.29 Step Three

Here is step three, the left hind. It is obvious that he's continuing forward: His right hind is in flight. I didn't have to touch him with my hand, but I did have it ready. Now, if Mr. Mellow was suddenly startled and started bolting backward, I would not try to stop him. I would just catch his lead rope on his way by, then calm him, and we would start over.

8.30 Butt-Bar Noise

With all four feet in, Mr. Mellow has found his treat in the manger. I raise the butt bar to get him used to the noise but I don't fasten it. There is plenty of length in this trailer for this very large horse.

Unloading from a Straight-Load — Demonstration of Goal

Sometimes getting a horse to back out of the trailer is more difficult than it was getting him to enter. In many trailers, you can't get next to the horse to properly cue him. So teaching the horse, in advance, to back up in-hand and "weaning" him from all cues except the voice command will help in unloading.

The first step off can be a tough one. Many horses initially become frightened when they can't feel the ground with a reaching hind leg, and they jump back into the trailer. That's why the obstacle work is so helpful. Slippery footing can add to a horse's fear, so never attempt to load or unload a horse on a patch of ice or mud unless you simply can't avoid it. If you are unloading without a ramp, make sure the horse's legs will not slip under the trailer sill as he backs out.

To avoid developing unloading anticipation in your horse, once you stop your vehicle, let him stand for about 5 minutes. Then open the window and let him stand a few minutes more. Now begin the unloading procedure. Never be in a hurry or allow your horse to be in a hurry.

When unloading from a straight-load, follow a very specific order of procedures to avoid mishap. First untie the lead rope or unfasten your horse from the trailer tie. Next, refasten the lead rope to the halter and drape the lead rope around the horse's neck. You can do this from inside or outside the trailer.

8.31 Open Doors

Open the door(s) and fasten them securely. Tell your horse "Whoa" if you think she needs some reinforcement. Here, I put my hand flat on Veteran's hindquarters to remind her not to step back until requested. Unfasten and move the butt bar, and be sure it is safely out of the way.

8.32 "Whoa"

Stand to the side for safety. If necessary, tell your horse "whoa" again. You want your horse to stand still until *you* give the command to back.

Require your horse to stand for 10 to 20 seconds.

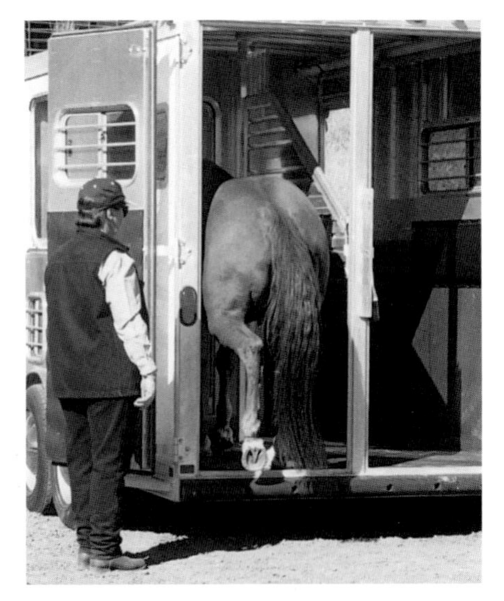

8.33 "Okay, Back"

Then, as you give the command "Baaaack," step to the side even a little more, just in case the horse backs out crooked.

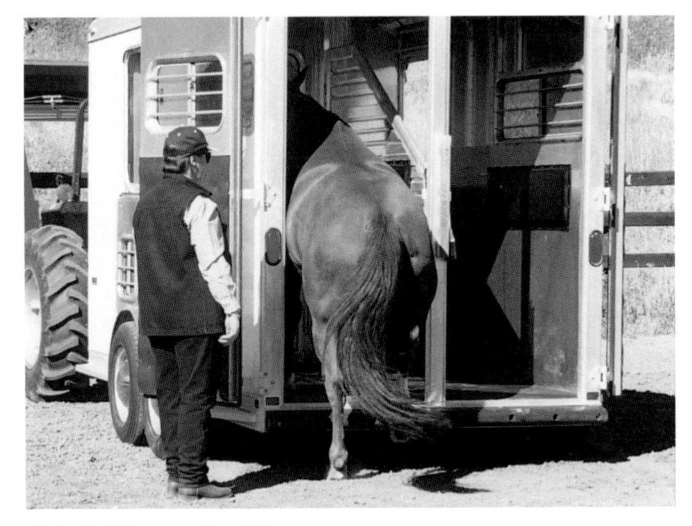

8.34 Take It Easy

As she takes that first step down, you can give her some positive reinforcement by saying "Goooood girl" in soothing tones so she knows she is doing well and to proceed calmly.

Continue speaking to her so she knows where you are. Unloading should be like loading — one step at a time, not a quick lunge backward. Keep your position and a distance of about an arm's length away.

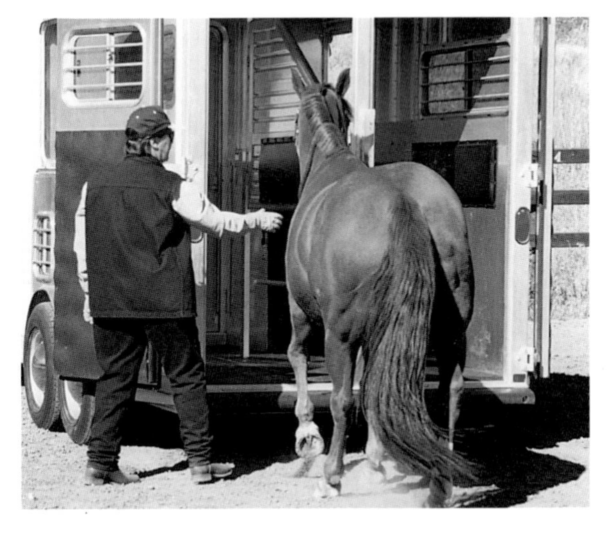

8.35 "Whoa"

When the last hoof is landing, say "Whoa" and reach over to take the lead rope.

Unloading from a Straight-Load — Training Session

Since I've found that more horses are reluctant to unload than load, the first time I unload a horse out of a straight-load, I usually give him some guidance and assistance. With the horse untied, the doors fixed securely open, and the butt bar unfastened and out of the way, I ask the horse to back as in the previous sequence. If he is "stuck," I enter the trailer via the empty stall or walk-through door from the tack room.

8.36 Bang to the Left

Mr. Mellow starts backing with a big swing of his hindquarters to the left, which causes him to bang into the trailer wall.

8.37 Bang to the Right

He overcompensates, and the next step results in a bang into the stall divider. This is common for the first unloading lesson, because the horse is, in effect, feeling his way out and learning the boundaries of his space.

8.38 Back Out Straight

The second time he is unloaded, he remembers that there are walls to avoid and backs out much straighter. Notice that in this first step down, when his head is the highest, he still has clearance in this 84-inch (213 cm) trailer. This is the proper-size trailer for this gelding.

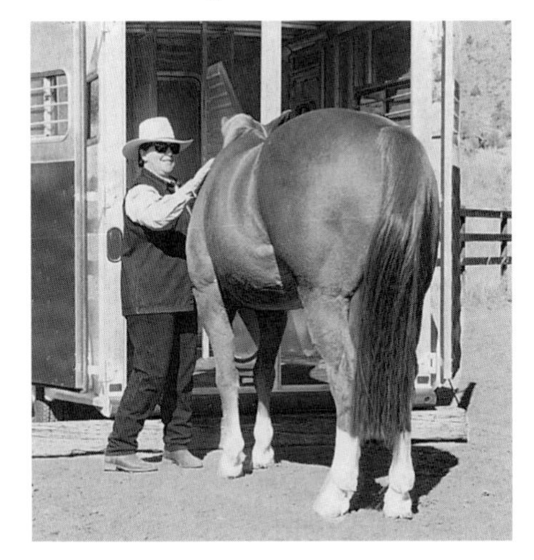

8.39 Good Job

We've had a great session, and I tell my horse that he has done well by giving him something he interprets as a reward — a scratch and a rub on his shoulder/withers area.

Tying

A horse's head should be secured in the trailer to stabilize it, restrict excessive movement, and to prevent horses from playing or fighting. In a straight-load, this is usually accomplished with a trailer tie, a 12- to 18-inch (31 to 46 cm) rope or web strap with a panic snap on one or both ends. The trailer tie is secured to a tie ring in the trailer.

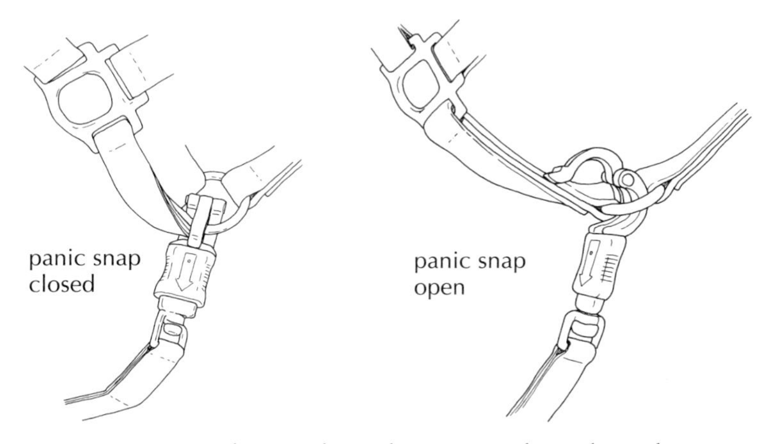

A panic snap is designed so that it can be released with one hand, even if the trailer tie is taut.

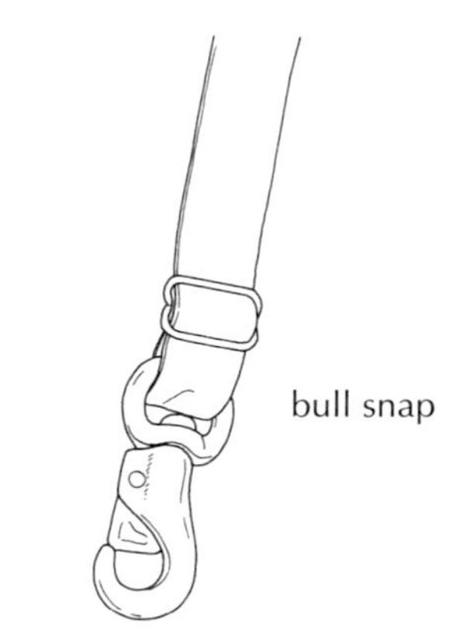

A normal bull snap is almost impossible to unfasten when the tie or rope is taut. Some trailer ties have a bull snap on one end and a panic snap on the other. This trailer tie features an adjustable slide so the tie can accommodate various trailers and horses of different heights.

8.40 Have Tie Rope Ready

In this slant-load trailer, the lead rope is tied to the overhead tie loop with a quick-release knot. The tail end of the rope has been dropped through the loop to keep the horse from untying the knot. One of the disadvantages of using a lead rope as a tie is the horse might chew on the rope in transit. The advantage of the lead rope is that in an emergency, you already have a rope on the horse — you just have to untie the knot. A trailer tie could be used here, instead, but if you needed to quickly lead a horse out of the trailer, you'd have to hold onto the short web tie, which isn't as safe as using a longer lead rope. This rope has been tied at a predetermined length that is ideal for the horse that will be hauled, making it easy to fasten the horse quickly when loading.

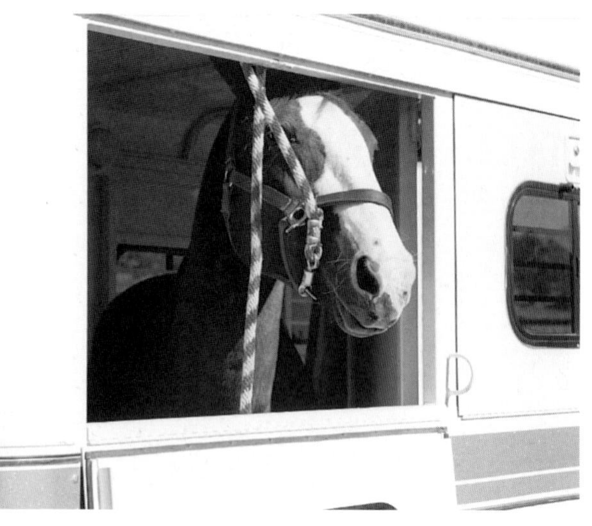

8.41 Too Short

Veteran is tied too short in this slant-load. She will have to hold up her head for the entire trip, which will make her neck very tired.

8.42 Too Long

Now she is tied too long. If she were inclined, she could reach all kinds of things on the outside of the trailer to chew on. Also, she could move back and forth so much inside the trailer that she could cause it to sway. She might also turn her head around inside the trailer, then get it wedged and panic.

8.43 Just Right

This is the ideal tie length for this slant-load. Veteran can assume a natural head position and not get in trouble. You can see the loop of the quick-release knot with the tail end of the rope dropped through the loop. Remember, though it is a good idea to open windows when you stop a trailer, *never* let a horse travel with his head out the window.

8.44 Too Short

Veteran is tied too short in this straight-load. She has very little choice as to where she can hold her head, and after a long trip her neck might be stiff and sore. Tying a horse so uncomfortably short could be cause for her to avoid loading the next time.

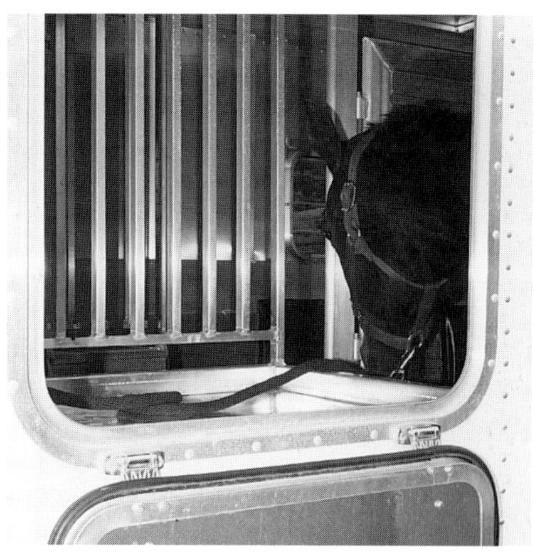

8.45 Too Long

Here, she's tied too long. The long rope gives her so much slack that she can put her head between the head grille and the center divider, where she could be injured or get in trouble with the horse in the next stall.

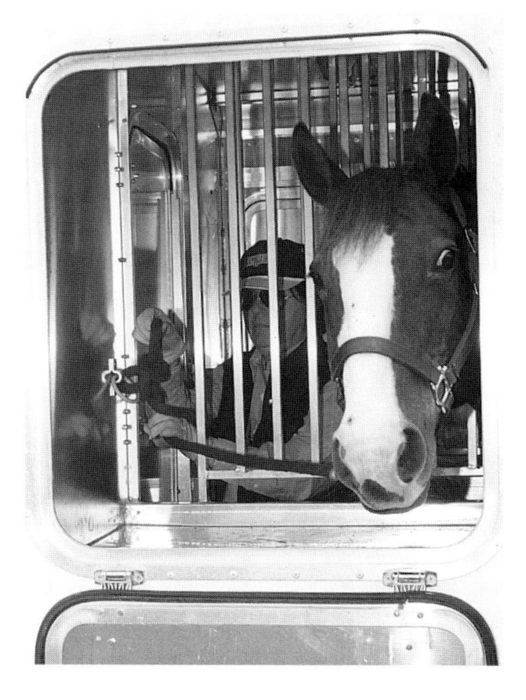

8.46 Just Right

Now, she is being tied just the right length. A quick-release knot is being tied to the manger ring.

9

✦ Tips, Cautions, and Story Problems ✦

Now that you've seen what to do and how, I'd like to give you a little food for thought for between training sessions.

The longer you're around horses and horse people, the more methods and tricks for loading you'll see. Although you'll never need to use tricks if you follow the progressive training program outlined in chapters 6 through 8, I'll talk about a few of them here. Also, I'd like to caution you about certain practices that are dangerous. Finally, I'm including a few common situations you might encounter when loading or unloading your horse, and I'll show you how to deal with them.

Though methods for loading vary among trainers, they should be based on the horse walking forward willingly on a slack line.

Use of Voice Commands

Physical aids, whether signals or discipline, should always be preceded by a verbal command such as "Walk on" or "Whoa." This gives the horse an opportunity to connect the verbal command with the physical pressure. In the future, it gives him a chance to respond to the verbal command to avoid the physical aid. Most horses are cooperative and seek to avoid physical correction; thus, they not only respond to voice commands but are able to discriminate and develop quite a repertoire as well. Clinicians who are demonstrating trailer loading often don't emphasize the use of voice commands because they are always talking to the audience. It would be difficult for a horse to distinguish a voice command amid the constant narration. Most clinicians emphasize the use of physical cues. But because most horse owners work alone with their animals and are not carrying on a conversation with an audience as they load, voice commands are not only appropriate but are also much better than physical cues in many cases. As the horse is learning the physical cues, couple the voice commands with them. Then eliminate the physical cues and use only voice commands. If a horse needs a tune-up, review the physical cues to reinforce the verbal commands.

Intensity of Aids

The intensity of a training signal, whether it is pressure at the horse's head or encouragement from behind, should be strong enough to get the desired response without eliciting fear or confusion. A too-strong signal may make a horse overreact dangerously or close his mind to future communication because he is afraid. A signal that is not firm enough can irritate a horse or may make him eventually ignore the signal altogether.

What to Do If the Horse Stops

If a horse stops to take an honest look at the trailer, let him. As long as it seems like he is inspecting or thinking about moving forward into the trailer, let him stand. If he quits considering, you need to initiate action.

When a horse stops out of stubbornness, make him walk on. If he won't walk forward because you are already near the trailer and he has not yet learned to get closer, you can move him to another area and review prompt in-hand work to reestablish forward movement, perhaps with a review of obstacles. Then return to the trailer.

When a horse needs encouragement to move forward, offer it first by the voice, such as "Walk on." If the horse doesn't respond, quickly follow the command with a tap from the long whip. Use the whip over the tail head or at the hip to be most safe. A whip aid below the hock tends to make a horse snap the leg up and often encourages him to move forward. However, if the horse is so inclined, it can also make him kick. Using a whip on the gaskin may also encourage a horse to kick, and might make him move sideways, out of loading position. Tapping a horse over the tail head will usually cause him to drop his croup, round his back, and flex his abdominals, which puts him in the position to climb into the trailer. A whip can be appropriate for a horse who has gotten "stuck" when it comes to forward movement. If the horse doesn't respond with a tap or two, it's time to go back to the in-hand and obstacle work until he will walk forward willingly. Overuse or misuse of the whip at the trailer will only make the horse resent the loading experience.

What to Do If the Horse Backs

If a horse backs up, resist the temptation to jerk on the halter. First test the thoroughness of your "Whoa" command. Horses that have been thoroughly conditioned will usually stop as soon as they hear the command. But sometimes the only thing you can do is follow a horse backward. Don't chase him backward but let him go as fast as he wants. Eventually, when he stops, get into a position where you can safely give him physical signals that will discourage backing. Poll pressure from the halter and a tap with a whip on the hindquarters are the most effective means to thwart backing. A forward and downward pull on a rope halter is the best way to deliver poll pressure.

CAUTION

If you use pressure on the noseband of a halter, especially with a chain, that actually tells the horse to lower his head and/or back up, which is not what you want when you are trying to get the horse to *stop* backing.

If an inexperienced horse steps into the trailer but then backs out immediately and stands quietly next to you, that's okay. Don't do anything. The horse is getting the feel of the trailer, and in this case, his backing out requires no discipline.

However, if a horse pulls you backward with him as he exits the trailer, he needs review of the in-hand work and stronger conditioning with "Walk on," "Whoa," and "Back." The temptation may be to use a chain to control a horse that misbehaves by backing. But remember, you can actually *cause* the horse to back by putting pressure on his nose with a chain; the harder you jerk on the chain, the more you are teaching him to back away from the pressure and from you and the trailer. What you need to do is reestablish forward movement.

What to Do If the Horse Circles in Front of You

When a horse tries to cut in front of you, pull back and push to the right on the lead rope. If you can, pop the horse on the nose, neck, or shoulder with the lead rope or poke him on the neck or shoulder with the butt end of your whip. Such a horse has clearly shown that he does not respect your personal space. Return to the personal space lesson as well as whoa, side pass, and whoa on a long line.

What to Do If the Horse Swerves His Hindquarters Sideways

If the horse's nose is still pointed toward the trailer, stop and have the horse "Whoa" on the long line. Organize yourself, center the horse by lowering his head, and then move him over so his body is straight. Then proceed to load (if that was your intention) or after a pause, walk away and approach the trailer again, this time straight.

Trailer Surprises

- **The horse comes barreling out of the trailer as soon as he is loaded.** Before you can fasten the butt bar, the horse rushes backward. *Never* stand directly behind a horse in a trailer, whether or not the butt bar is fastened. Rushing backward is a dangerous habit that is difficult to cure, so do not let the horse ever think about rushing out of a trailer. Use the one-step-at-a-time method for both loading and unloading, as outlined in chapters 8 and 9. If you have purchased a horse with this habit, seek the help of a professional trainer.

- **The horse rears and puts his front feet in the manger or gets his front feet over a tie rope or a breast bar.** If a tie rope has a panic snap and/or a quick-release knot, you can free the horse. However, it is very difficult to get a horse's front feet out of a manger, especially if the manger has a lip. The horse is standing on his hind legs and the front legs are fully extended, yet the front legs need to come *up* to come off the manger. The horse cannot do this himself unless he thrashes violently and rears again, which often results in him banging his head in the trailer or flipping over backward, and that could be paralyzing. Rearing and getting hung up usually only happens with a small horse in a large trailer and is most commonly associated with yearlings that have not had adequate leading and tying lessons. The more thoroughly you prepare a horse with groundwork, the less likely it is that this will happen. Refer to tying guidelines in chapters 6 and 8.

- **The horse loads into a straight-load, then tries to turn around in the narrow stall and gets wedged.** Again, this is a situation that is most common with a young or small horse or pony in a large trailer. The only sure way to prevent this is to lead the small horse into the trailer so you have control of the head. This way, the horse can't turn around. Once a horse is wedged, it is difficult to get him to bring his head back around to straighten, but if the horse continues to struggle to turn, he could damage his shoulders or hips. This is one of the main reasons always to tie a horse in a trailer.

- **The horse paws or kicks in the trailer.** First be sure the horse has learned to stand tied to a hitch rail or post for several hours without pawing. This may take a period of a week or more to accomplish, tying the horse every day. If the horse stands quietly tied outside but paws in the trailer, check the following common reasons for pawing and fix the problem. There might be wasps or hornets in the trailer; find and remove any nests. The horse might be covered with irritating flies on his legs; use fly spray. The trailer might be too enclosed and warm; open all windows and vents. The horse might be trying to get at feed that is on the floor; pick it up and give him some hay in a hay bag. The horse might be playing with another horse or another horse might be pestering him; separate them. The horse might be too crowded; give him more room.

What About Tranquilizers?

Should you use a tranquilizer for your horse to minimize his mental stress about loading and traveling? In general, the answer is no. Using a chemical restraint in place of proper training is not a good policy. Sedated horses have a more difficult time keeping their balance while traveling, have a lower resistance to illness en route, and will require substantial time upon arrival to recover their faculties. In addition, some horses have a hyper-reaction to tranquilizers and may become more sensitive to lights and noises while traveling.

What Other Types of Restraint Are Suitable for Trailering?

Some neutraceuticals (nutritional supplements that have a pharmaological effect) have been developed that tend to calm a nervous horse. The best advice here is to ask your veterinarian whether there is anything in the product that you are considering that could harm your horse. If not, plan to "test" your horse's reaction to the product several times before you load him so you know how he will react. If the product calms the horse but he is still coordinated, cooperative, and alert, it might be useful for a nervous traveler.

What About Using Feed for Loading?

Trailer loading and unloading should be approached as any other training lesson — commands or signals from the trainer to obtain desired responses. Do not teach a horse to load by bribery with feed: If you ever have to load a horse when you don't have any feed or if the horse isn't real hungry, you're sunk. There is a big distinction between feed as a bribe and a treat as a reward once the horse has loaded. Leaving a treat in the manger is an excellent way for a horse to find a reward upon compliance. Never feed treats from your hand.

Is Physical Force Appropriate?

The physical aids and cues outlined in chapters 6 through 8 are not considered physical force. Scare tactics and brute strength, however, are considered physical force and have no place in training a horse to load. Physical force includes two strong people "boosting" or lifting a horse's hindquarters into a trailer while another person keeps the lead rope taut through the manger. Pulley ropes and butt ropes are also considered physical force. Some horses who have been forced to load with ropes routinely refuse to load until the ropes appear, then they just jump in. This proves that horses have a very strong power of association and so are trainable. Take the time and work on the right lesson and the horse will retain good habits for life.

Test Ride

It is helpful to take a horse that is new to trailering on some short practice rides with a well-mannered companion. This must be a horse that the newcomer is familiar with and does not fear or want to fight or play with. With the manger full of fresh hay and a friendly, seasoned traveler alongside, the horse will be more likely to relax and enjoy the association with the trailer. Take it slow. Avoid big bumps, sudden starts and stops, and fast turns.

DON'TS

- Don't try to physically push your horse into a trailer.
- Don't use physical force to *make* a horse load in a trailer.
- Don't stand directly behind a horse in a trailer, whether or not the butt bar is fastened.
- Don't use an escape door unless you are well aware of the risks, have proper footwear, and know the horse very well.
- Don't pull a horse's tail to ask him to unload.
- Don't let a horse rush out of a trailer.
- Don't let a horse lunge into a trailer.

Slowing Down the Horse That Leaps into a Trailer

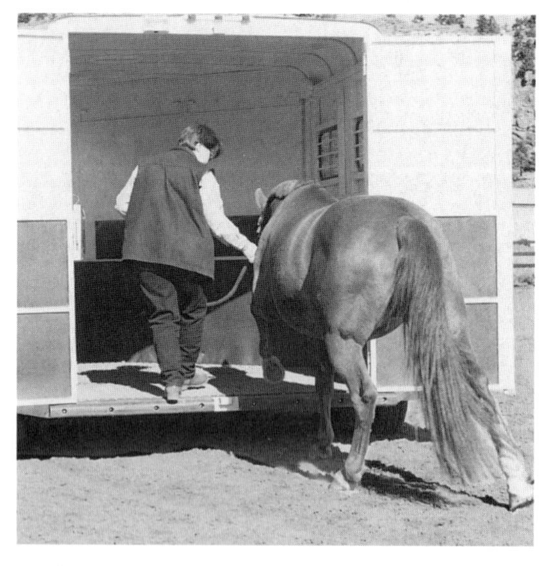

9.1 Start to Load

If you have a horse that leaps or lunges dangerously into a trailer, lead her into the trailer with her off side close to the far trailer wall to contain her. I'm using Veteran here who is hardly a rusher or leaper, but she will demonstrate the technique. You can use a similar lesson with a horse that unloads too quickly.

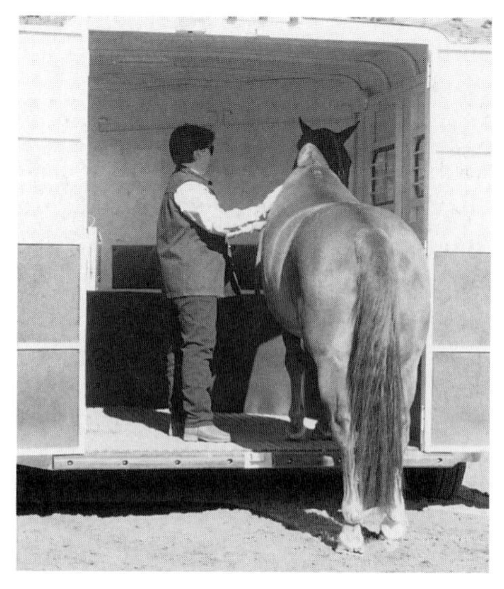

9.2 Stop Halfway In

When the horse has two feet in the trailer, stop. Let her know that it is safe to stop halfway in. Give her a pat. Wait sufficient time until the anticipation has lessened. Continue walking in calmly and slowly. If you can't stop the horse at the midway point, go back to the in-hand lessons and work the obstacles one step at a time.

Avoid Bad Habits by Insisting on Good Manners Early in Training

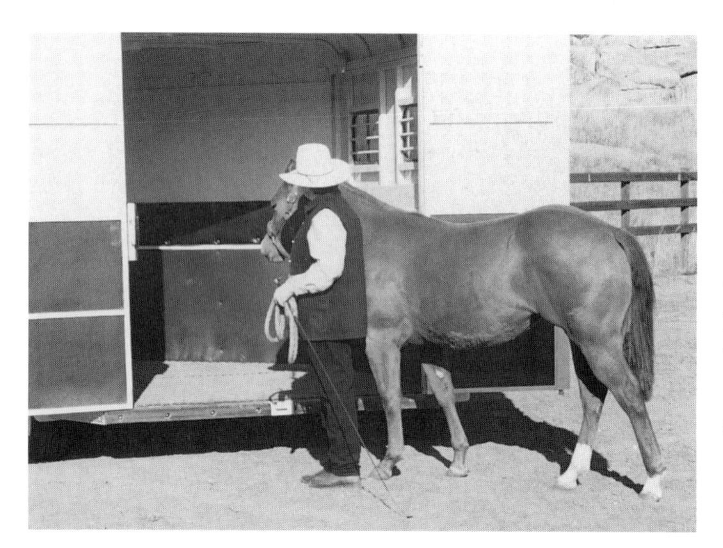

9.3 Walk By

Rookie is having her first loading session. She is brought up to the trailer opening and allowed to look in. She doesn't appear too concerned, but her body is not straight enough for loading.

I want to walk her by the open trailer anyway, as she had concerns about walking past the closed trailer in an earlier lesson. In that lesson, I walked her past on a slack line. She didn't rush or crowd although she did pay attention to what was on her right.

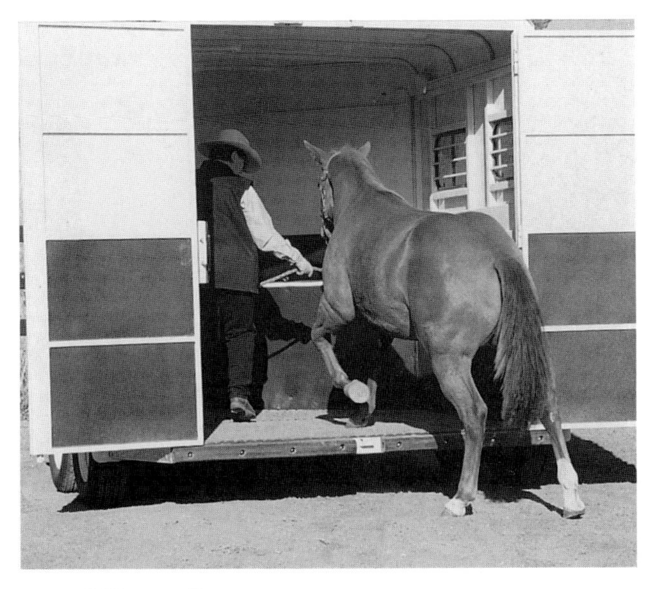

9.4 Walk In

The next time we approach the trailer, she keeps her body straight. I ask her to walk in and she does so calmly, one step at a time.

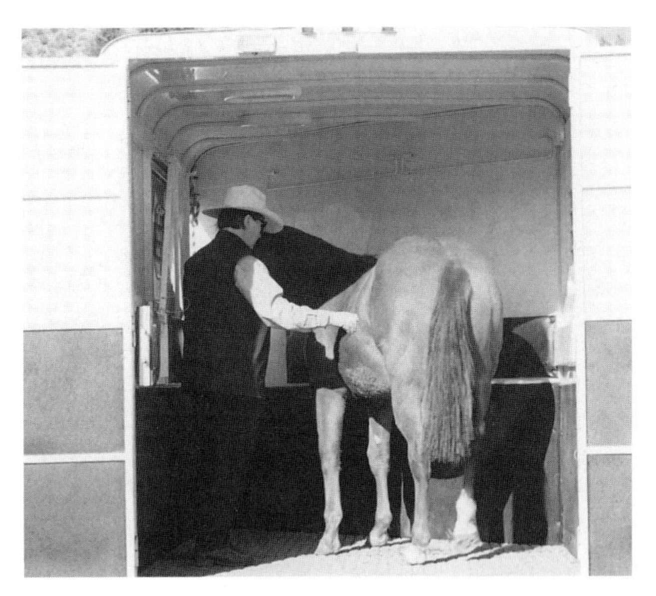

9.5 Turn Around

When Rookie has all four feet in the trailer, I pause for a moment, give her verbal praise, and let her settle. After she settles, I begin turning her, asking for alternate steps of turn on the forehand and turn on the hindquarter.

Even though Rookie is still somewhat small, when she gets to the midpoint of the turn, I sense she is getting wound up like a spring from all the confinement and wants to bolt out the door. So as I finish the turn, I quickly but smoothly switch my hands on the lead rope so that I am holding it with my right hand when we complete the turn.

9.6 Urge to Bolt

Just as I suspected, Rookie wants to bolt from the trailer. Although the chain on her halter was unnecessary for all maneuvers before and after this moment, I'm glad it's in its engaged position now. With one smooth move, along with the "Whoa" voice command, her forward movement is stopped.

Avoid Bad Habits by Insisting on Good Manners Early in Training (continued)

9.7 Exit Slowly

After pausing 5 seconds, I relax the tension somewhat on the lead rope but still retain control. She exits calmly one step at a time.

9.8 Relaxed Inside

The next time I load her and turn her around, Rookie automatically stops at the end of the turn and waits for further commands. This deserves a great amount of praise and rubbing. She has done very well and will keep these habits for life.

PART III
TRAVELING

When you travel with your horse, the number one priority is for both of you to arrive safely. For peace of mind, develop thorough and safe trip preparation and trailering habits.

Develop and follow maintenance programs for your truck and trailer as outlined in chapters 1 and 4.

Prepare and plan before every trip using checklists you develop in chapters 10 and 11.

Use practical, effective horse-care procedures en route as outlined in chapter 12.

Follow through with post-travel care in chapter 14 for the horse, truck, and trailer so they are ready for the next trip.

10

✦ Getting the Rig Ready for a Trip ✦

The most important part of getting ready for a trip is to be sure the driver is physically and mentally prepared. If you are the driver, take care of all trip preparation well before bedtime so you can get a good night's sleep. You will be sharing the road with small cars, pedestrians, large trucks, other horse trailers, motorcycles, bikes, and the occasional deer or antelope: You must be alert for anything. If you're well rested and organized, you'll be a safer driver.

Have your truck and trailer serviced regularly, so they are always ready at a moment's notice if you need to take your horse to the vet or evacuate in case of emergency.

Make a spare set of keys for ignition, glove box, tonneau or camper, spare tire lock, trailer hitch lock, tack room doors, trailer windows, and anything else that has a lock on it.

Pre-Trip Truck and Trailer Check

- Tire pressure
- Wheel lugs
- Hitch
- Lights, running, turn signals, brakes
- Trailer brakes
- Emergency trailer breakaway system

Keep a complete set of spare keys in a safe, accessible place in the towing vehicle. In addition, put a master vehicle key in a magnetic box in a protected place on the exterior of your tow vehicle.

When you prepare your rig for a trip, perform a pre-trip safety check, then pack the truck, hook up your trailer, and pack it before you head to the barn to get your horse ready. It's a good idea to perform the pre-trip rig check a week before you go — this way, if you find any problems, you have time to fix them. Then give your whole rig a recheck the day of your trip.

Tire Pressure

Check the pressure on all truck and trailer tires when the tires are cold (photo 10.1). It is normal for a warm tire to run with a slightly higher air pressure. Look for the specification, such as 50 psi, on the sidewall of your tires and be sure to inflate them to that pressure. Running tires at lower pressure can cause overheating and even blowouts. Routinely inspect both your truck and trailer tires for irregular tread wear, bulges, defects, and weather checking. You don't want to find any surprises that cause you unnecessary delays before or during a trip.

Check the tire pressure on the trailer spare. Be sure it is located in a safe place where it cannot be stolen. If it is on the outside of the vehicle, make sure it has a lock on it, and know exactly where the key to the lock is located.

10.1 Be sure all tires are inflated to their specifications.

Tire Pressure Tip

Pressure specifications are listed on the sidewall of your tires and will likely be different for your truck and trailer tires. For each type of tire you are using, write the tire size, the vehicle that it is on, and the correct tire pressure on a 3 x 5 card. Keep the card in your glove box or visor organizer along with a tire gauge. Sometimes mud or poor lighting will make it difficult for you to read the designation on the sidewall, so you will really appreciate the note card.

Check the tire pressure on the truck spare. This may require you to crawl under the vehicle. Once you get under your truck to take this reading, you'll understand why it is important to stow the spare so that the stem side of the wheel is facing the ground.

Wheel Lugs

Check to see that wheel lug nuts are tight. Many trailer manufacturers emphasize this in their owner's manuals and via decals placed on the trailer fender (see page 139). You'll need to use a torque wrench to be certain you're meeting the manufacturer's specifications. (See the tire-changing sequence in chapter 12 for more information.)

Hooking Up Your Trailer

When you are breaking in your coupler, at first it might be reluctant to slip onto the ball. You can help it glide on by rubbing grease or a bar of soap over the ball. Soap is not as messy as grease, and this is a good way to use up soap scraps. A well-lubricated ball allows the coupler to move freely over it and reduces ball and coupler wear from friction. Make sure the nut securing the ball to the ball mount is tight.

Once you get the coupler resting on the ball, if you need to, give the coupler a kick to close the hinged portion so that the collar of the hitch slides over it. (See chapters 1 and 2 for more on the hitch.)

10.2 Grease or soap the ball.

Breakaway Mechanism

See chapter 3 for a description of this safety feature.

10.3 Remove Pin

To fasten the cable of the breakaway mechanism to your truck, remove the breakaway pin from the switch. This requires a strong pull. Wrap the wire cable around your gloved left hand and grab the cable close to the coupler with your right. Give a sharp jerk straight backward.

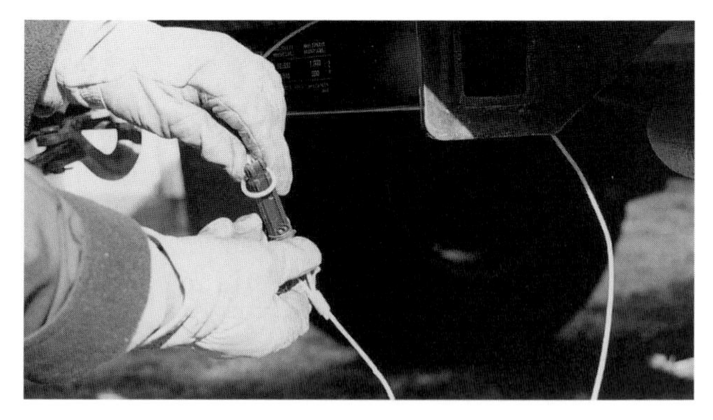

10.4 Run Cable Around Truck Frame

Run the cable around the frame on your truck and pass the pin through the loop on the other end of the cable. Pull the pin end of the cable to take the slack out of the cable. Plug the pin back into the switch.

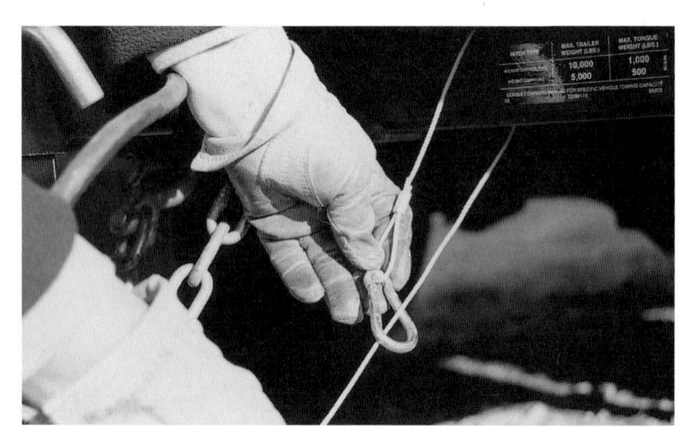

10.5 Customize

To customize your breakaway cable so that each time you hook up you don't have to pull the pin, attach a small climber's carabiner to the end loop of the cable. When you want to attach the cable to the truck frame, just pass the end around the frame and snap the carabiner onto the cable. A spring-loaded gate on the side of the carabiner holds the carabiner closed. Carabiners are made from stainless steel; the one in the photo is 2 inches (5 cm) long.

Check the voltage of the breakaway battery several times a year and trickle-charge when necessary. (See chapter 4 for more information.)

10.6 Hitch and Breakaway Ready

Here is the breakaway cable running from its loop around the frame to its point of attachment at the switch box on the side of the trailer's tongue. Take note of some other significant things in this photo: The cable is slightly longer than the chains, which means that the cable won't be activated until the chains break or become disconnected. The battery that runs this mechanism is in a waterproof plastic box affixed to the trailer tongue. The keepers on the chain hooks prevent the hooks from jumping out of their attachment holes. Also note that a carabiner has been added to the end of the breakaway cable to facilitate hooking up.

Hitch Safety

The trailer coupler and the truck's receiver are essential for your safety as well as that of your horse. Develop a method of hooking up your trailer and do things in the same order every time. Then, after you have finished the process, go over each step again. Before you drive off, again go over everything related to the coupler, receiver, breakaway cable, and electric brakes. Make sure that the hitch is not cracked or rusted, that it has no loose parts, that the coupler is securely seated over the ball, and that the locking mechanism is engaged. Whenever you plug or unplug the electrical wire, do it by holding onto the metal plug, not the wire sheath. It won't take more than a few times yanking a plug out by the wire for you to start having electrical problems.

Check the inside of the trailer for such things as hornet nests in the ceiling and within the vents. Often they are difficult to spot. Hornets also like to place their nests under fenders and under the floor of the trailer. It only takes one encounter with a wasp to teach a horse to refuse to load or to become a very nervous traveler. It's a good idea to carry a can of bee or wasp spray along with you during nesting season in your area.

Examine all nooks and crannies of the trailer and tack room for signs of insect or rodent damage or presence. Because you cleaned the trailer after its last use, there should be no nesting materials left behind to invite rodents. But sometimes they take up residence rather quickly, especially in the fall. Besides leftover hay in mangers, mice will gravitate toward any feed or blankets you left in the tack room.

10.7 Clean Bed
Clean out the bed of your truck so that debris doesn't blow around and fly at the trailer. This is essential if you are towing a gooseneck so that the hitch, chains, and cables will operate freely and don't get hung up. If the breakaway cable is yanked free by a loose object, you and your horses could get a bad scare at the least, or be in a serious accident.

10.8 Raise Jack Wheel
When you get ready to move your trailer, raise the hitch jack so the wheel is at its full height. If the wheel is not all the way up, when you go over uneven ground or even just in or out of a driveway, the wheel could drag on the ground. If it hits hard enough, it could pop the hitch loose. Some wheels ride very low to the ground even when they have been raised to full height, and some trailer manufacturers make the hitch wheels removable for this reason. If removable, stow the wheel where you can find it when you get to your destination.

Loading the Rig

Load your rig with all emergency items and creature comforts that are appropriate for the type of trip you're taking. If you are just transporting your horse a few miles up the road to the veterinary clinic for vaccinations, you won't pack as you would for a four-day trail ride or a two-day horse show. Each trip has its own special gear requirements. On all trips, however, make sure you are equipped for emergencies. Use the items and suggestions in this and the following chapters to help develop your own lists.

Truck Ready Checklist

- ❑ Engine oil
- ❑ Power-steering fluid
- ❑ Transmission fluid
- ❑ Brake fluid
- ❑ Coolant in both radiator and recovery reservoir
- ❑ Windshield washer fluid
- ❑ Wiper blades
- ❑ Air filter (especially if you drive in dusty conditions)
- ❑ Battery terminals clean of corrosion
- ❑ Fill tanks with gas, either beforehand or at your earliest convenience

Emergency Equipment

When you break down along the highway, you'll be glad you have emergency equipment along.

Locate the jack and know how to use it. Learn where the various portions of your jack and crank are stowed. Know where the spare tire is and how to get it down and to refasten its carrier. Bring along a can of tire sealant or a small air tank to repair punctures, either of which might give you enough of a temporary fix to get to a repair shop. (See chapter 12 for tire-changing procedures.)

Quick fixes for common problems include extra motor oil, a radiator hose, a fan belt, fuses, and bulbs.

Large black garbage bags can be used as a ground cover when you're changing a tire or working on a vehicle or horse. They make good emergency rain ponchos — cut out head and arm holes and use the drawstring to snug it up around your waist. They're good for emergency water storage (best within a bucket), collecting manure, gathering roadside garbage, and putting over the saddle if caught riding in the rain.

Small zipper-lock bags are good for grain storage; to keep small items dry; as emergency gloves for dirty jobs; to store screws, bolts, and parts when making repairs; and for emergency hand, foot, or hoof soak (use several and tape around top).

Duct tape is useful for hoof protection in case of a lost shoe; to repair blanket or trailer boot tears, radiator hoses, and damaged sneakers or boots; as emergency additional bandage material; for covering up sharp edges; and to tape garbage bags closed.

Bring along electrical tape for wire repairs to truck and trailer, to keep trailer parts from rattling, and as additional bandage material.

Tools for Changing a Flat Tire

If your trailer has hubcaps, carry a tool to remove them. Here are three choices, starting from the left: a crowbar, which can also be used as a pry bar to get the tire into place; a heavy screwdriver; and a straight tire iron with tapered end. On the right is a four-way (star) wrench, which will likely have sockets to fit both truck and trailer lugs. The star design makes it easy to apply a lot of balanced force to loosen and tighten wheel lugs. Use a dab of nail polish or paint to designate which arm of the star wrench has the socket to fit which lugs.

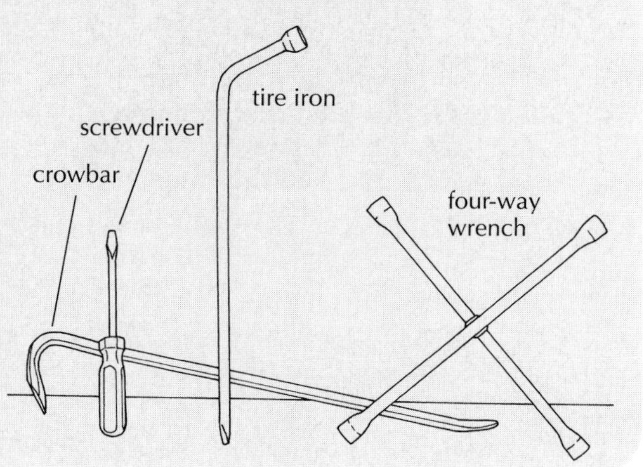

10.9 Tire-Change Helpers
Tire-changing essentials include (from center, counterclockwise) a tongue wheel chock, flat boards on which to place jack when on soft ground, Trailer Aid Jack (E-Quest, Inc., Medford, NJ), a truck jack, Jiffy Jack (Rule Steel, Inc., Caldwell, ID), and blocks to block the wheels if you need to unhitch. The Trailer Aid Jack is suitable for trailers up to 20,000 pounds (9,072 kg); one size fits trailers of all types. The Jiffy Jack is available for horse trailers with torsion bars or leaf or slipper springs. Choose one that is specifically made for the size trailer you have. For example, this is a size small for a torsion bar, two- to three-horse trailer with a loaded trailer weight under 8,000 pounds (3,629 kg). It lifts the other axle 4½ inches (11 cm), enough to remove a flat and replace it with a fully inflated tire. Jiffy Jacks are available in medium, large, and extra-large for loaded trailer weights up to 20,000 pounds (9,072 kg). (See photo 12.10 for a ramp in use.)

Take along a plastic tarp to lie on if you need to crawl under your truck or trailer or to kneel on when changing a tire; an old shirt or jacket to keep your traveling clothes clean; and gloves, rags, and old socks to use with messy breakdown chores.

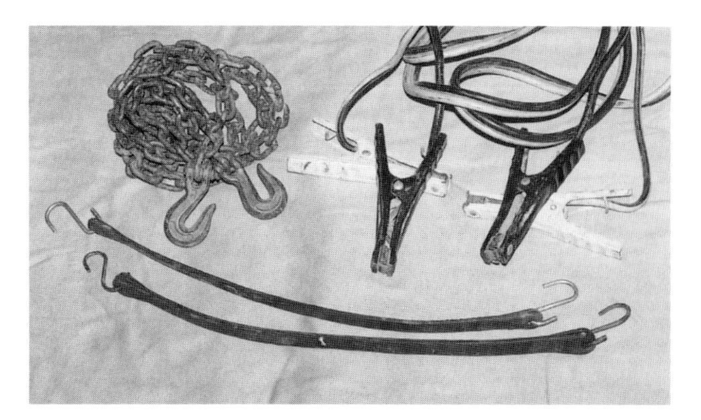

10.10 Roadside Aids
To help you with your disabled vehicle, from top left: a tow chain, to move the rig off the road or help you get unstuck; jumper cables, to boost a dead battery; and extra bungee cords, to secure broken parts such as dividers and grilles, to hold open doors on windy days, and to secure objects in the tack room, on the roof rack, or in the pickup bed.

10.11 Hand Tools
Whether or not you are proficient with tools, it's a good idea to have them along in case someone who stops to help has tool know-how. This prepackaged kit includes a set of open-end wrenches, sockets, screwdrivers, pliers, and a safety knife. In addition, pack a can of WD-40 lubricant and (in bottom row) a handy pocket multi-tool, additional pliers, Vise-Grips, and a hammer.

10.12 Emergency Essentials
In addition to the items in this photo, if you plan to travel during snowy months, consider carrying a shovel, sand, and chains. Here, clockwise from the left, are triangles, fire extinguisher, flashlights, extra batteries, and flares. Carry three collapsible, reflective, weighted triangles to alert oncoming vehicles that you are stopped ahead. (Refer to chapter 12 for more about their use.) The 11-ounce (261 mL) fire extinguisher contains potassium bicarbonate. It is rated 2-B:C, which means it is appropriate for grease, gasoline, and electrical fires. Be sure it is charged. The large flashlight is waterproof and has a clear chamber that allows the light to shine out like a lantern when the flashlight is set down. It also has a handle so you can hang it from your belt or wrist or from the open hood of your truck. The small flashlight also has a loop and is small enough that you can hold it in your teeth if you need both hands to set the jack under the trailer, for example, to change a flat tire in the dark. A battery-powered headlamp would also be helpful. Always carry extra, fresh batteries because night breakdowns often last longer than the set you have in your flashlight. Use flares at night to alert oncoming traffic when you are stopped along the road.

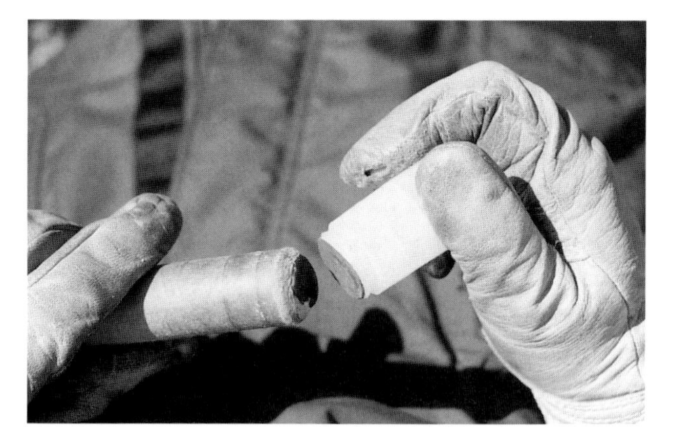

10.13 Lighting a Flare
To light a flare, remove the protective plastic end, which is also the strike-plate cap. Then strike the black igniter material on the end of the flare with the rough strike plate of the plastic cap. Sparks fly, so do this away from your eyes and any flammable clothing. If there has been an accident, be careful to keep flares away from any fuel that has leaked onto the ground.

Prepare for Detours and Alternate Routes

Know where you are going and be prepared for differing state regulations. Have exact directions to your destination and plot your trip on a map so you won't all of a sudden encounter a highway regulation that would prohibit you from proceeding!

Equipping the Rig

Keep a copy of your truck registration, trailer registration, and proof of insurance in your towing vehicle at all times. Some states may require other vehicle documents as well.

I like to carry a halter and lead rope in the cab of my truck in case I have to quickly catch a loose horse or get a horse out of a trailer. During an accident, a horse's halter or lead rope might break and the trailer tack room could be locked or inaccessible.

For long trips and those in remote places, you may have to do some impromptu camping or hoofing it if you break down. To make your overnight stay in the horse trailer or truck more comfortable, pack a sleeping bag, toilet paper, and moist towelettes. For cold weather, bring along insulated winter boots, a warm hat, and warm gloves. For summer, a broad-brimmed hat, comfortable walking shoes, sunscreen, and lip balm will come in handy.

To keep the driver and copilot fueled and operating properly, pack nutritious, high-energy snacks with a relatively long shelf life, such as Power Bars and nuts; water, tea or juice, and a hot beverage; prescription drugs; and prescription sunglasses and clear glasses, or contact lenses.

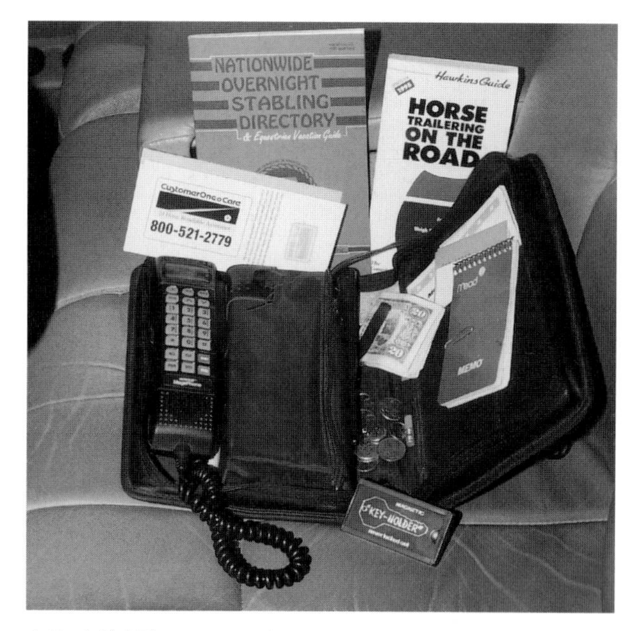

10.15 Travel Accessories

Take along a map, horse motel guide, a cell phone (or CB radio if you will be in remote areas), your 24-hour emergency assist card, important phone numbers, cash (change for tolls and phone calls; folding money for emergency towing in some cases), and, at bottom, a magnetic key case with a spare key. Find a safe metal resting spot on the underside of your truck or trailer to secure the key case.

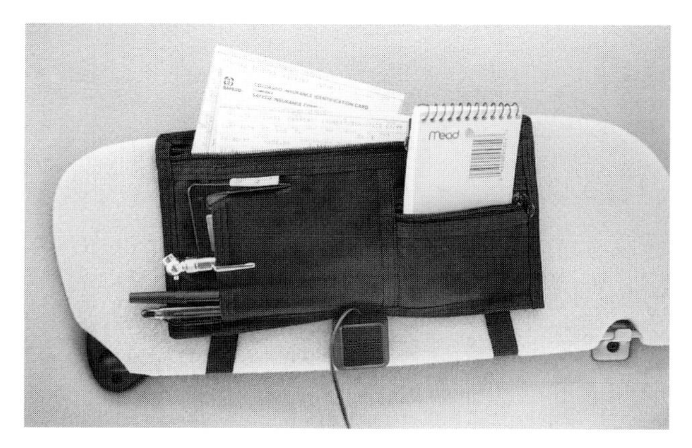

10.14 Vehicle Documents

This handy visor organizer has all paperwork at the driver's fingertips, plus a tire gauge, notebook, and other small items. In the notebook, have a list of important phone numbers, including a way to contact a veterinarian or friend who can take responsibility for your horses if you become incapacitated during the trip.

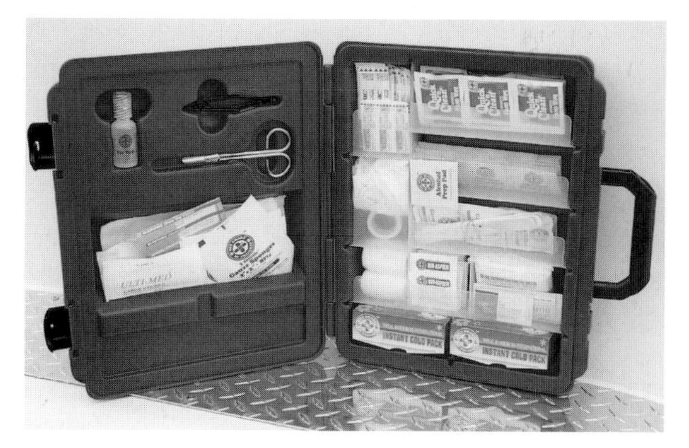

10.16 Human First-Aid Kit

A prepackaged first-aid kit (for people) contains basic minor-wound-care materials such as antiseptic, antibiotic ointment, adhesive strips, and aspirin. Create your own kit from scratch or start with a commercial kit and beef it up to include a better pair of scissors, some more-serious bandaging materials, your brand of anti-inflammatory medication, and latex gloves.

Packing Horse Supplies

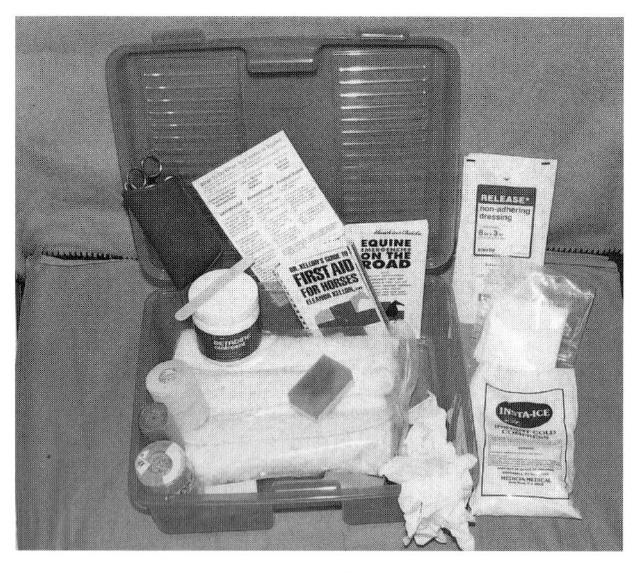

10.17 Horse First-Aid Kit

A first-aid kit for horses should include a good supply of bandaging materials such as nonstick gauze pads, conforming gauze, padding (you can use disposable diapers), self-adhering stretch bandaging material, and elastic adhesive tape. Also pack a clean set of leg wraps to go over any bandages you might need to apply. Carry some povidone-iodine (Betadine) wound wash, saline eyewash, scissors, triple antibiotic ointment, betadine ointment, latex gloves, a thermometer, and a stethoscope, too. Ask your veterinarian to recommend and supply you with any additional items he or she feels you would be capable of administering.

Travel Tidily

Be certain to pack a muck bucket, plastic rake, aluminum scoop shovel, plastic fork, and broom. The cleaner you keep the trailer while in transit, the less bedding you'll use, the cleaner your horse's leg boots will be, and the more comfortable it will be for you and your horse. Don't leave manure and urine to begin decomposing with the bedding; the ammonia buildup could irritate your horse's respiratory tract.

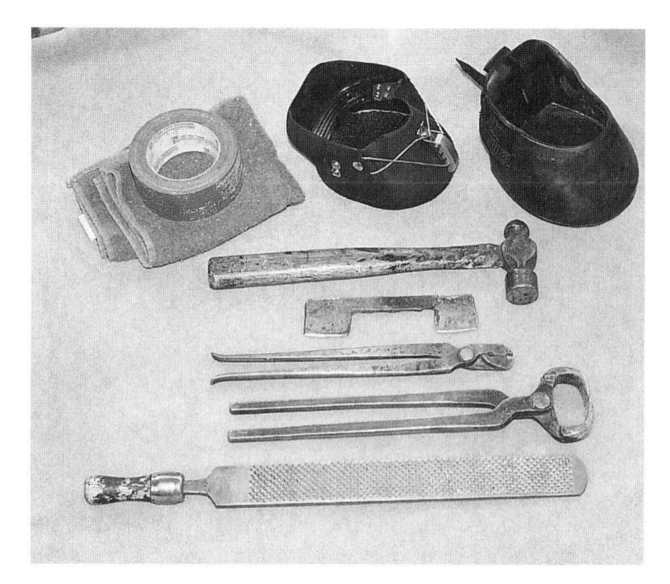

10.18 Horseshoe Emergency Kit

A hoof kit should include, from bottom: rasp, pull offs, creased nail puller, clinch cutter, hammer, tape, cloth, and hoof boot.

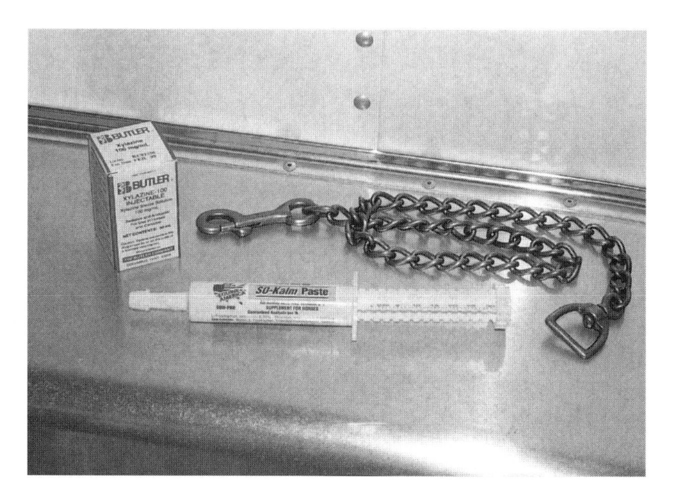

10.19 Restraint

Choose the method of restraint that you are most comfortable with: drugs, neutraceuticals, or a halter chain. (See chapters 6 and 9 for more information on these options.) Restraint may be necessary in case of an accident or if you have to unload your horse in a strange or dangerous area.

10.20 Feed

Pack the feed that your horse will require while traveling, at your destination, and for the way home. For a two-day trip, perhaps a bale of hay and premeasured grain rations for each feeding marked A.M. and P.M. are all you'll need. Zipper-lock bags make handy storage pouches for the feed. In the event that you break down and your horse requires an extra feeding, you can routinely pack more for an emergency or pack a pail of large hay cubes or wafers (foreground). In addition, bring along a small salt brick (in front of the grain on the bale of hay). The small pieces left over from a large salt block come in handy for traveling.

10.21 Electrolytes

Electrolytes are ionic substances such as sodium, potassium, calcium, magnesium, and chloride that are necessary for many body functions. Depending on each individual horse's metabolism and the type of hay you feed, you may choose to give your horse electrolytes when he is traveling or working hard. Some examples are powdered Gatorade, powdered electrolyte mix, and paste electrolytes. Don't use electrolytes indiscriminately. If he is not accustomed to the taste, and you heavily lace your horse's drinking water with them, he may go off water, which would intensify dehydration problems.

10.22 Water

Take along water for drinking and washing. The white pail contains a garbage bag filled with water. The bag is sealed securely with a twister tie. If you position the bucket in the truck or trailer where it won't tip over, the bag will prevent water loss from sloshing.

10.23 Water Storage Caddie

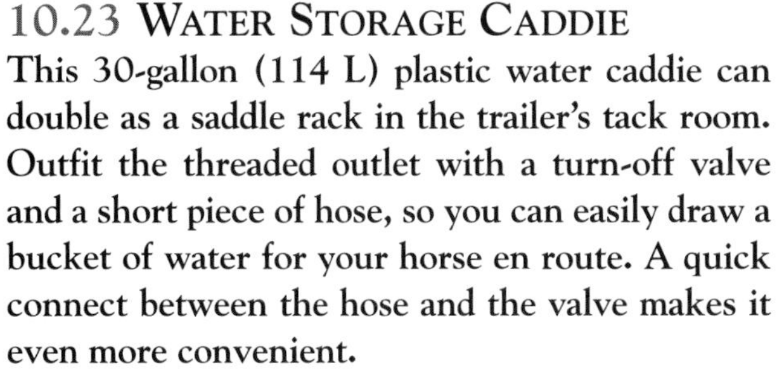

This 30-gallon (114 L) plastic water caddie can double as a saddle rack in the trailer's tack room. Outfit the threaded outlet with a turn-off valve and a short piece of hose, so you can easily draw a bucket of water for your horse en route. A quick connect between the hose and the valve makes it even more convenient.

Getting the Trailer Ready for the Horse

In most cases, it is best to haul a horse without bedding in the trailer. However, if you plan to be traveling for four hours or more, bed the floor with sawdust or shavings to encourage urination. Use only very clean sawdust or shavings. If the bedding is dusty, your horse will be breathing dust for the entire trip, and you could be unloading a coughing horse at your destination. If the bedding is only slightly dusty, give it a sprinkle of water. However, if it is very dusty, don't use it at all. When hauling mares, just bed the rear portion of the trailer to catch the manure and urine. When hauling geldings or stallions, you might wish to bed the entire floor, better to absorb the urine, which is directed more forward.

Before you load a horse, open the overhead vents. On warm days, open all the windows as well. It is easier to reach the overhead vents before you load a horse than it is to crawl over a sweating horse to open them en route. Most vents open two ways. For air to flow in, such as during hot weather, open the vent so the air rushes in the opening from the front. If you want to draw warm, moist air out of the trailer during cooler weather, and you don't want a breeze inside, open the vent so that the air is drawn out of the trailer toward the rear. Which windows you open will depend on the temperature, weather, and if horses are clipped or blanketed. Horses do well in the cold but are easily stressed by extreme heat or a cold draft.

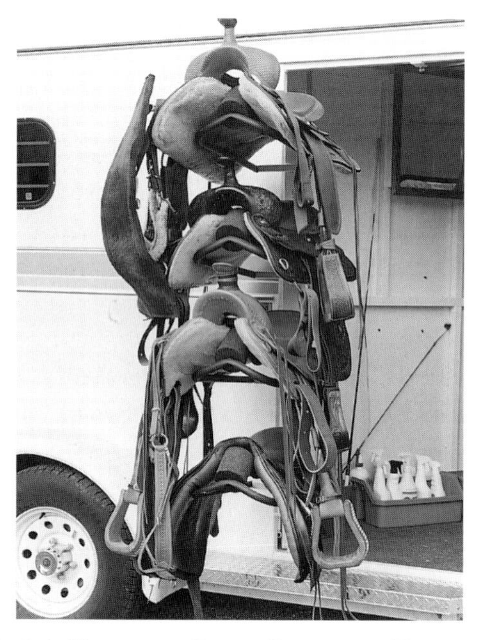

10.24 Swing-Out Saddle Rack

A well-appointed tack room will help keep your items organized and orderly. Features might include a swing-out saddle rack, a door organizer, a boot box, medicine chest, bridle hooks, storage bins, and clothes rod. This swing-out saddle rack holds four saddles. Note the medicine chest with mirror on the interior wall. Some cabinets require auxiliary latches to ensure that they don't pop open en route.

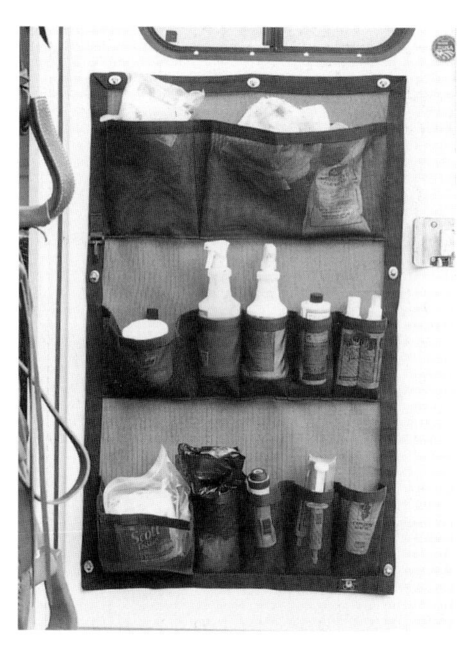

10.25 Door Organizer

This door organizer holds a multitude of items that you want to have within arm's reach when you're working on a horse tied to the side of the trailer. You also can use a closet shoe organizer for this purpose. With a clear or mesh organizer, you will be able to see what you have inside each pouch. This provides an easy-to-reach place to store brushes, bandages, medications, a flashlight, and other emergency supplies.

10.26 Hay Feeder

For short trips, feeding is not necessary for nutrition. Feeding hay can occupy a horse, though, and chewing will help relax him. If you are going to feed in a slant-load or a straight-load without a manger, use a safe feeder. Mount it high enough that a horse can't get a hoof caught in it. A hay net is not a good trailer feeder choice — you must put it up high enough that the horse can't possibly get a foot caught in it, but the higher you place a hay net, the more dust and flakes the horse inhales as he eats. With any feeder, consider soaking the hay to reduce dust and prevent respiratory stress. Soak, drain well, and then load the hay bag. Be aware that damp hay molds quickly in warm weather. (See photos 3.25, 10.27, and 11.1 for more on feeder choices.)

10.27 Manger Feeding

If you are hauling in a straight-load with a manger, load the hay loose in the manger or tie it in place in a hay net. When using a hay net, ensure that it is tied very securely and that when it is empty, the net cannot in any way entangle the horse. A hay net does make hay soaking convenient, and helps contain the feed so it is not spread all over the trailer, but the risks of using the net often outweigh the convenience. Note the flush, slam latch on the adjacent dressing room door.

Final Checks

Check running lights, turn signals, brake lights, emergency flashers, and brakes to be sure they are working on both the truck and the trailer. Have your traveling companion watch while you operate each light.

Check electric brakes and adjust them if necessary. This should have been done in your annual maintenance (see chapter 4), but you may have to adjust your controller from time to time according to the load. Refer to the instructions that came with the controller for adjustment procedures. If you've adjusted your controller with the trailer empty, you will likely need to make an adjustment when towing a loaded trailer.

Adjust the side mirrors so that you can see alongside the truck and trailer. Check the final adjustments when you are out on a straight stretch of road.

Always leave a written itinerary with your family, business partner, neighbor, or a friend, so in case of emergency, there will be information to help locate you.

11

✦ GETTING YOUR HORSE READY ✦

Minimize travel stress for your horse. Whatever type of conveyance a horse is traveling in, he can experience harmful mental or physical stress if he is not properly prepared.

Horses greatly prefer wide open spaces. When a horse obeys your command and loads in a trailer, he is exhibiting great trust in you. You need to accept full responsibilty of his care and be aware of things that could harm him.

Dampness, dust and other airborne debris, fluctuating temperatures, and drafts are what lower a horse's resistance and set him up for illness. A horse needs to regularly clear his respiratory tract by lowering his head and blowing. Often a horse is blanketed too warmly for a trip. This causes him to sweat and he then catches a chill. It's a good idea to open overhead vents to allow warm, moist air to escape from the trailer without causing a draft. And it is generally more appropriate to blanket a horse in layers rather than with one heavy blanket. Then check him frequently and remove layers as needed.

Flavoring Water

Prevent dehydration by preparing your horse for water changes. Every horse's tastes and tolerances are different regarding small variations in water. Some horses will drink water from almost any source; others are so finicky that you will have to devise a means of disguising water from an unfamiliar source. It is almost impossible to carry enough water to satisfy a horse's requirements for more than a day or so. Water is simply too heavy and bulky to haul very much of it. At home, flavor your horse's water ahead of time. Using trial and error, you can find what "flavoring" works for your horse. Start with a small amount of something that has a pleasant but distinctive smell and taste, such as apple juice. Add just a few ounces to your horse's water bucket at home for a week before traveling. Work up to a full 8-ounce (189 mL) cup per bucket if you anticipate encountering very odd water on your trip. If you see that your horse is drinking well with the apple juice at home, use it to flavor the water on your trips to keep him drinking and hydrated. One of the leading causes of travel colic is dehydration, so keep a close eye on your horse's water intake before and during all trips. Other substances that can be used to mask water include vinegar, molasses, flavored powdered electrolytes (see photo 10.21), oil of peppermint or wintergreen, vanilla extract (just 1 or 2 drops per bucket), a sprinkling of Kool-Aid or Jell-O, and, in an emergency, a splash of soda pop. Often the sweetness will encourage a horse to drink. But ascertain what your horse likes *before* a trip.

Feeding En Route

Feed the same hay to the traveling horse that he is accustomed to at home (photo 11.1). It is usually safest to decrease a horse's grain ration several days

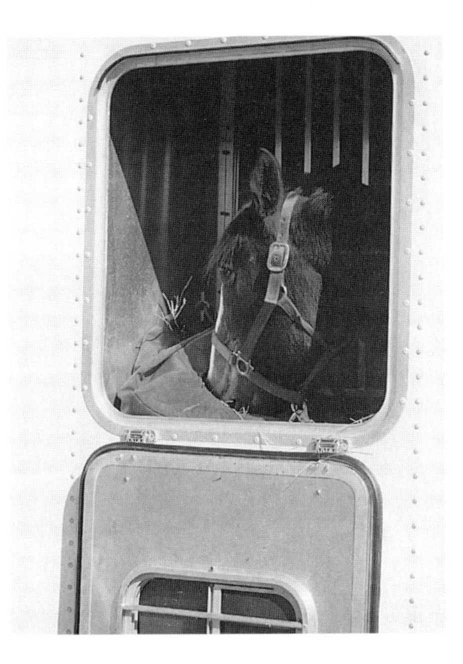

11.1 **Feed small amounts of absolutely dust-free hay en route (see page 125).**

before a long trip and eliminate grain altogether while traveling. On your veterinarian's recommendation, your horse may receive a dose of mineral oil or a bran mash the night before departure. This will help keep the contents of the horse's intestinal tract soft and moving.

Health Papers

Depending on the states you will be traveling through and the requirements at your destination, you will need various health papers for your horse. (See the appendix for a list of health requirements by state.)

Some horse shows and events have strict requirements and won't let you on the grounds until your paperwork complies. Some regional livestock health issues arise suddenly. This means you should regularly check local or regional horse publications for additional requirements. For example, if you are bringing in a horse from an area that has had a recent outbreak of vesicular stomatitis, you may need to have the horse examined before entering the show grounds. In some instances, regional problems have closed down shows altogether. You are responsible for knowing the regulations, and it is wise to stay informed about any new situations and outbreaks.

Generally you will be required to present for each horse:

1. A recent health certificate (usually within 30 days of departure). This document, also called a Certificate of Veterinary Inspection, or CVI, is prepared by your veterinarian after he or she has examined your horse.
2. A recent (usually within 6 or 12 months, depending on the state) negative Coggins test for Equine Infectious Anemia (EIA). A horse can be a carrier of EIA without exhibiting symptoms of the disease. A laboratory test (Coggins) is necessary to determine whether a horse has EIA antibodies in his blood. The presence of antibodies is considered a "positive" result. It shows that the horse has been exposed to and is a potential transmitter of the disease. If your horse is negative, he is not a carrier and poses no threat of spreading the disease.
3. A permanent ID card or Brand Card or temporary brand inspection papers for the trip. This will depend on the states you are traveling through.
4. A copy of the horse's registration papers. Many breed shows and events require that you present a copy of the horse's registration papers when you check in at the show.
5. In some instances, you will be required to submit a proof of vaccination for certain diseases.

In states that require brand inspection (for the most part, only those west of the Mississippi River), an inspector from the State Brand Board (usually listed under the State Department of Agriculture) will need to look over your horse to verify his color, markings, age, and sex according to his registration papers or bill of sale. Brand clearance generally is necessary only if you travel outside the state or if the horse is transported 75 miles (121 km) or more within the state. If you plan to do a lot of traveling, get a permanent brand card for your horse from the State Brand Board.

Getting the Horse Ready

Prior to traveling, any horse should be in good health and good flesh, and current as to all necessary vaccination and deworming schedules.

If the horse is nearly due for boosters, give them to him at least two weeks before departure. This will allow his immune system to react to the vaccines and raise his blood titer for the various diseases he may encounter when traveling or by coming in contact with other horses. Although annual vaccinations for sleeping sickness and tetanus are important, it is crucial for the traveling horse to be well protected from the respiratory viruses that cause influenza and rhinopneumonitis and also from strangles bacteria. It may be recommended that horses receive boosters two or more times per year against these diseases. Ask your veterinarian for his or her advice.

Be sure the horse has had proper training so that he will load calmly and without hesitation or gimmicks anytime, anywhere.

Outfit your horse with:

- Shipping boots
- Tail wrap
- Fly mask, to keep pests and debris from harming his eyes (optional)
- Head bumper (optional)
- Traveling sheet or blanket (optional)

Shipping Boots or Leg Wraps?

Your horse should wear shipping boots every time he travels in the trailer. Shipping wraps, however, are another matter. Properly applied, these offer added support for a long trip, but poorly applied trailer bandages can cause problems. Bandages that come undone can entangle a horse's legs. And worse, opponents to shipping wraps say that any pressure on the legs is detrimental. They interfere with circulation, especially as the horse is relatively inactive during trailering.

Use high-quality shipping boots to prevent injury to the coronary band and bulbs of the heels. In order to be effective, the boots should have thick padding and a durable exterior. They should reach at least to the midpoint of the hoof, cover the bulbs of the heels, and have a reinforced heel plate. Protection of the knee and hock is also desirable. Fitting the boots low means that they will contact the ground, get dirty, become scuffed, and possibly tear if the horse steps on them, but the coronary band and bulbs of the heel are precisely where the protection is required. Shipping boots also keep the horse's legs clean while in transit.

Shipping boots rarely cause pressure: They are usually contoured and thickly lined with fleece or foam, which makes them difficult to put on incorrectly or too tight. Most shipping boots also provide extra protection to the knees and hocks. Some shipping boots and shipping bandage combinations are quite hot and thus inappropriate for long trips. If you use boots or bandages, make certain that they are secure. Accustom a horse to wearing boots or wraps ahead of time or they might cause him to fuss, stomp, or kick.

Tail Wraps

Wrap the horse's tail to protect it from rubbing as he leans on the butt bar, door, or trailer wall during sudden stops, rapid acceleration, and when traveling uphill. Cotton bandages are safe for wrapping tails. Because they are slippery, though, they are often found on the trailer floor upon arrival. Good results are obtained for short trips with an elastic bandage or stretch crepe bandaging tape applied with light to moderate, even tension. Notice where your horse's tail hits the butt bar in the trailer and be sure to wrap his tail to that level and below, so when he leans on the butt bar he doesn't rub off tail hair.

Training Tip

Habituate your horse to shipping boots, a tail wrap, a fly mask, a head bumper, and a traveling sheet before you load him in the trailer. You want to protect him with these items, not add more stress or complications to his loading and traveling.

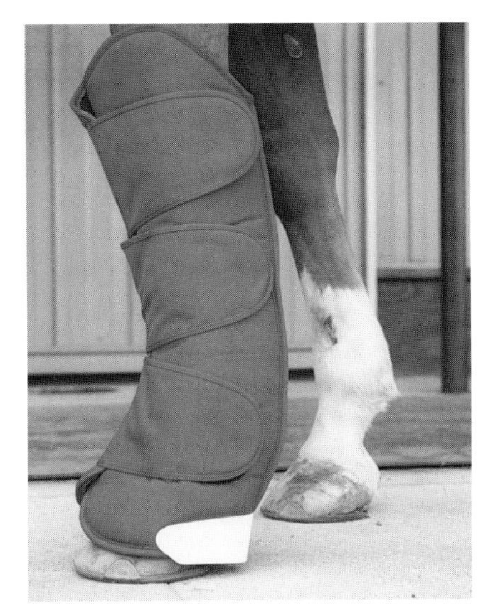

11.2 Shipping Boot
This shipping boot has three wide hook-and-loop closures and a plastic heel shield. Note that the opposite leg has an injury from scrambling when the horse was not wearing protective boots. (See photos 7.19–7.23 and 8.15 for more on shipping boots.)

11.3 Poor Wraps
This horse obviously did some moving around: His leg wraps are in shreds. The wraps should cover the coronary band — these stop short, leaving the coronet and heel bulbs exposed.

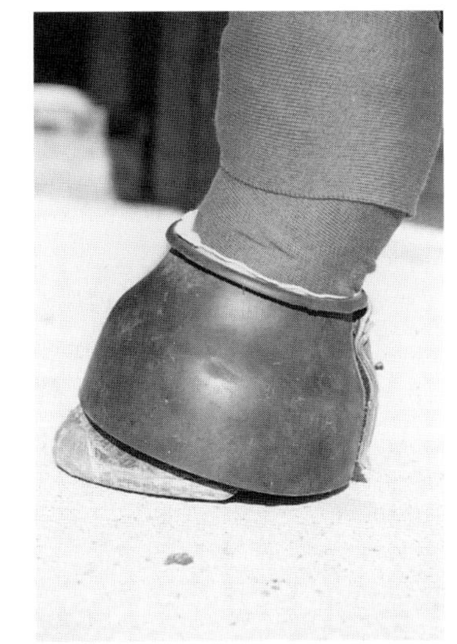

11.4 Bell Boot
To protect a horse's coronet and bulbs and also to protect shipping wraps, use a bell boot over the bottom of the wrap in this manner. You may have to buy bell boots one size larger than usual so they will fit over the thick bandage.

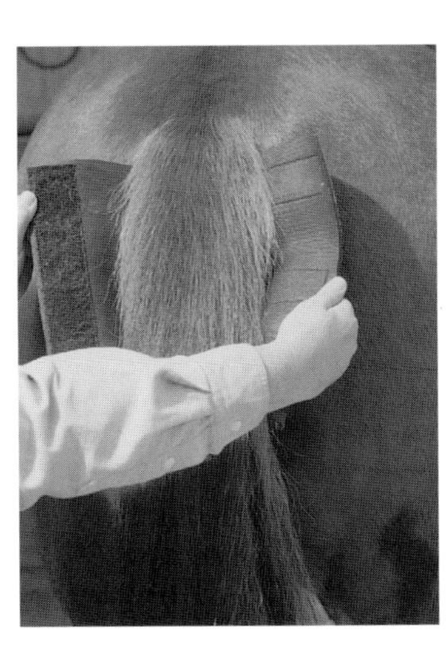

11.5 Position Tail Wrap
This simple neoprene tail wrap has a rubbery inside surface that grips the hair, holding it in place. Place the wrap as high on the tail as possible. You may need to lift the tail to get the wrap situated high enough under the top of the horse's tail.

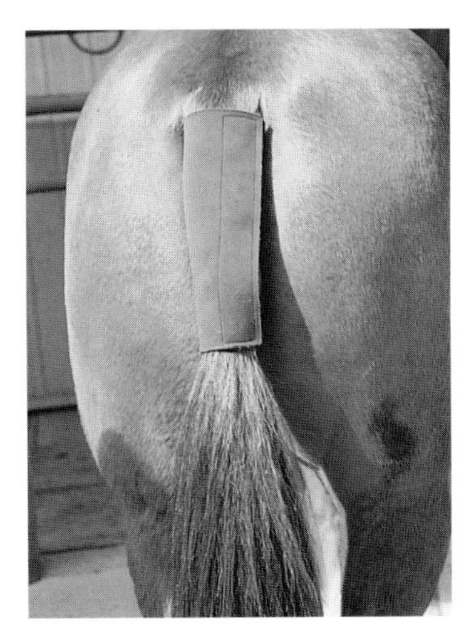

11.6 Fasten Closure
Close the neoprene with snug pressure and fasten the hook-and-loop closure. Braid or wrap the lower part of the tail to further protect it if desired (see photo 3.13).

Tail-Wrap Caution

Use light to moderate, even tension as you wrap the tail. Nonslip rubberized tail wraps really stay put, but it is easy to apply them too tightly. If a tail wrap is too tight or left on too long, it will cut off the circulation to the dock (the upper flesh and bone portion of the tail), and the tail hairs will fall out. Often sloughed hair will grow back in a lighter color, white or blonde.

Blanket or Sheet?

During warm weather, use a light sheet to keep dust from settling on your horse's clean coat (photo 11.7).

If it is very cold, you can put a winter blanket on your horse for traveling, but it is better to cover the horse with two light layers rather than one heavy blanket. That way you can remove or add a layer to fit changes in temperature. In some well-insulated two-horse trailers, the heat generated by a pair of horses can create a situation that is too warm and moist for their health. Keep overhead vents open at all times, monitor body heat, and open windows when appropriate.

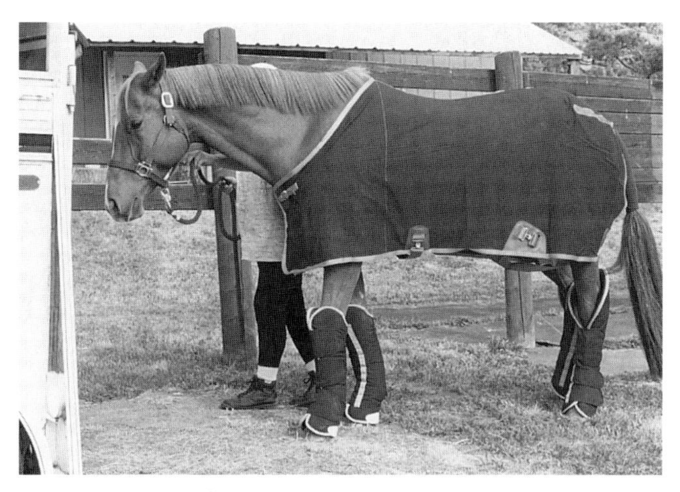

11.7 **Traveling sheet**

Head Bumper

A head bumper, which is a heavy leather poll cap that fits between the horse's ears and protects the poll, is extra protection when trailering tall horses. A horse that has a tendency to raise his head excessively during loading or unloading would benefit from a head bumper. The bottom line, though, is that you should be using a trailer that is tall enough for the horse.

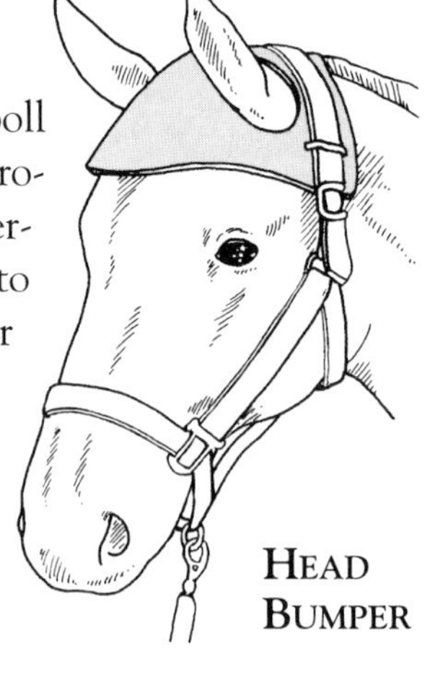

HEAD BUMPER

Fly Mask

A fly mask makes a horse more comfortable during warm weather when all of the windows are open or if you are hauling him in a stock trailer (photo 11.8). The mesh of the mask allows the horse to see but protects his eyes from airborne debris, like hay chaff.

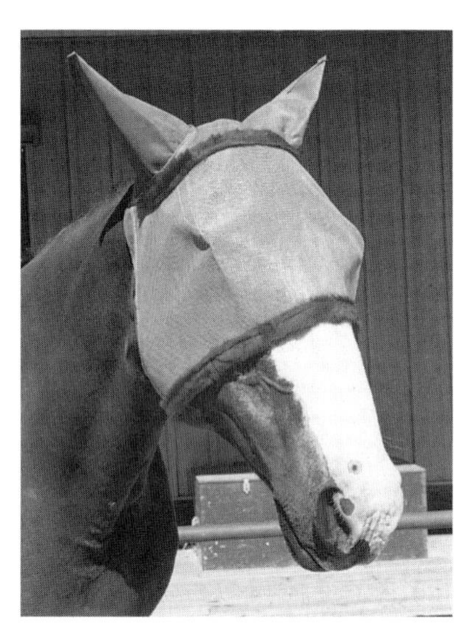

11.8 **Fly mask**

Load 'em Up

If you are using a straight-load trailer and hauling one horse, load him on the left side, so he will ride up on the crown of the road. If you put him on the right side, his weight will have a tendency to pull the trailer off onto the shoulder.

If you are hauling two horses in a straight-load, put the heavier horse on the left and the lighter horse on the right, for the same reason.

In a slant-load, always load a single horse in the front stall to keep the weight at the front of the trailer. Never put a single horse in the rear stall or you might be in for some trailer sway. When hauling two horses in a slant-load, put the heavier horse in the front stall.

12 ✦ En Route ✦

Drive carefully, according to the advice given in chapter 5. Always allow yourself more traveling time than seems necessary. Don't use car-driving times to estimate how long it will take you to get somewhere. Know your route before you set out. Pack a detailed map with explicit instructions to your final destination. Locate rest stops and places where you can get help along the way. If you don't belong to a 24-hour roadside-assist program, consider becoming a member.

Be sure all passengers are buckled up. Children belong in the back seat, and in safety seats if appropriate.

Trailering can be very hard on a horse's muscles, bones, joints, ligaments, and tendons. Rough roads, long miles, inexperienced or inconsiderate drivers, inadequate trailer suspension, poor floor mats, and improperly maintained tires all can cause unnecessary wear and tear on your horse. To help ensure that a horse arrives at the destination refreshed rather than fatigued, make the trailer as safe and comfortable as possible.

Horse trailers can be hot. Haul during cool hours: morning, evening, and night. Avoid parking a trailer in the hot sun for extended periods. Keep vents and grilled windows open whenever possible.

Trailer Check

Shortly after you leave home, stop for a trailer check. As long as you can pull off the road safely, anywhere will do. In fact, plan to stop routinely every 100 miles (161 km) or so. Give yourself a stretch while you walk around the rig. Check the tires and the hitch and take a look at the horses. During such a stop, you may be able to detect a problem, such as a low tire or a hot brake, early — before it becomes a disaster.

During one of your stops along the way, you may wish to clean the manure from the back of the trailer. But be careful. Whenever you open a trailer door alongside a highway, you're taking a chance — a passing vehicle could catch the door or your horse could think you're unloading him and try to bolt. If you practice brief stops at home, as part of your horse's training, you should be okay.

Weigh Stations

Do you need to stop at a weigh station when you are pulling a horse trailer? Each state has its own regulations about this. Some states have no permanent weigh stations but will set up temporary checks along the highway. Any state patrolman can stop and weigh your vehicle with a portable scale.

Your state's regulations related to weigh stations might read "All vehicles pulling horse trailers must stop. Brand and health papers will be checked," or "All vehicles transporting livestock must stop at the state port of entry," or "All commercial vehicles must stop and all commercial and noncommercial vehicles over 16,000 pounds (7,257 kg) GVWR [Gross Vehicle Weight Rating] must stop." There are many variations. You must find out the current law for your state. (See *Hawkins Guide to Horse Trailering on the Road.*)

Rest Stop

When it is time for a more substantial rest break, perhaps you can find an area where you can use the facilities and check on your horse.

Whenever possible, whether at a rest area or your final destination, try to park the trailer in the shade. It will make your horse's traveling home so much more comfortable.

Checking Your Passenger

If your horse is wearing a sheet or blanket, slip your hand under it. A horse that feels wet is either too hot or has broken into a nervous sweat. Depending on the situation, change the blanket, unload the horse, or put a cooler on him to help him dry gradually. Check and adjust vents and windows. A lightweight plastic mounting block comes in handy for trailer-side checks.

Hydration. Check your horse's level of hydration using the pinch test (photo 12.3). Constantly be on the lookout for signs of dehydration. Practice performing this test at home, so that you know what normal is for your horse. With your thumb and forefinger, grasp some skin on the horse's neck. Lift it up into a tent. Let go of the skin and see how long it takes to return to its flat position on the neck. A return in 1 or 2 seconds is normal, but skin that is dry and slow to recover means the horse is dehydrating; a standing tent for 5 seconds or more means serious trouble. If you suspect a problem, take the horse's temperature, pulse, and respiration and capillary refill time. Know which values require you to seek the help of a veterinarian.

12.1 Park with the Trucks
Generally you will be directed to a special area where semi trucks park and the parking spaces are bigger.

12.2 No Room with Trucks
If the truck spaces are all filled, you might have to park across some of the passenger-car spaces. When the rest area is not busy with motorists, this should be okay. Realize, however, that you are taking four or five spaces for your rig. This rest area was filled with trucks but there was only one passenger car in the auto section.

Open the trailer windows so your horse can fill up on fresh air.

Water. Offer your horse water every 2 to 3 hours. The most common cause of dehydration during travel is a horse going "off water"; that is, refusing to drink. Horses detect differences in the smell and taste of water and sometimes won't drink water away from home. A horse should drink 5 to 6 gallons (19 to 23 L) per day — more when the weather is hot and humid and when the horse is competing or lactating.

Carry as much home water as is practical, such as in a tack room water caddy (see photos 10.22 and 10.23). If you don't have a way to carry home water or if your home water supply runs out, you might need to flavor the new water (see photo 10.21). Be sure to accustom the horse to the flavoring in his water at home for several days before the trip.

If you are concerned that the horse is not drinking properly and have prior experience or advice from your veterinarian, administer paste electrolytes en route. This may stimulate your horse's thirst reflex.

Urination. You can encourage a horse to urinate en route by bedding the trailer stall deeply with sawdust. Always remove urine-soaked bedding to prevent irritation to the horse's respiratory tract from ammonia fumes.

Some horses that are reluctant to urinate on board will urinate readily when unloaded along a grassy roadside or in a turnout pen. Be aware that if you take your horse for a hand walk or to urinate in a rest area or alongside a highway, the vegetation there may have been sprayed with herbicides for weed control. Although you might like to let your horse have a good neck stretch and a few relaxing bites of grass, this could result in colic.

Keep your horse on his regular feeding schedule. In general, clean, grass hay is the safest traveling ration, but feed your horse the type of hay he is accustomed to.

Take care of yourself during the rest stops, too. Use the opportunity to walk around and stretch and get some fresh air.

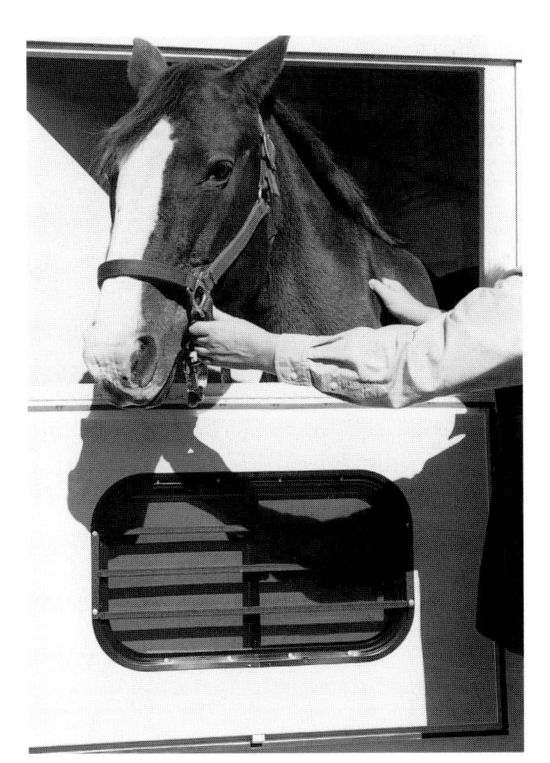

12.3 Pinch Test
Monitor your horse's level of hydration with the pinch test.

12.4 Water Frequently
Offer your horse a drink every 2 to 3 hours.

Checking Your Rig

Look at all trailer and truck tires. Know what looks normal. Check the tire pressure on any that look low. Feel tires and wheels. Know what is normal heat and how much heat signals brake problems. If the wheels are smoking, or you can detect the acrid smell of hot brake pads, or the wheels are so hot that you can barely touch them, you definitely have brake problems. Call for help at once.

12.5 **Check wheels and tires.**

12.6 **Check the hitch, chains, electrical connection, and the breakaway cable. Check all door latches.**

Emergency Stop

If it becomes necessary to pull over for an emergency, stay calm and carefully bring your vehicle to a stop. Often you don't have a choice of where you can stop and pull over when your radiator overheats, or if you have a flat tire, but always look ahead to see if there is a wide shoulder or a roomy place where you can pull off the road. Put the transmission in park (automatic) or first gear (standard), turn off the ignition, set the parking brake, and turn on your emergency flashers.

Follow the recommendations in your state's driver's manual for indicating a disabled vehicle. Here are some general guidelines. Choose a safe plan that fits the specific situation you are in. Position the first triangle about 10 feet (3 m) behind your truck with its reflective side facing the traffic that is approaching your rig from the rear. Place the second triangle 100 feet (30 m) from the first. You can estimate 100 feet (30 m) by taking forty to fifty steps. Depending on your situation, place a third triangle 100 feet (30 m) farther or at the front of the vehicle facing oncoming traffic.

Triangles are good day and night, but as an alternative at night, you can use flares. (See photos 10.12 and 10.13.)

Engine Overheating

One of the most common tow-vehicle problems is an overheated engine. This generally occurs when you're pulling up a long grade. If the temperature gauge is climbing dangerously high, you still may be able to prevent overheating. Pull off the road and let the engine idle until the gauge returns to the normal range. If the front of the radiator is covered with bugs, brush them off as best you can so air can flow through the radiator. *Stay away from the inside of the radiator!* That's where the fan is.

If the engine is so hot that steam is leaking from under the radiator cap, leave the engine running and trickle water over the radiator to cool it. If you turn off the engine, the water stops circulating through the engine and it will get even hotter, forcing most of the coolant out through the radiator cap and overflow hose. *Never* pour water directly on the engine.

If you don't have water with you (shame!) and the temperature gauge doesn't start to fall within a minute or two after you've pulled over, you may be dangerously low on coolant already. Now you'll have to shut off the engine for about 45 minutes to let it cool. Then, with a thick rag and/or a glove, unscrew the radiator cap (you have to push down as you unscrew). If steam starts shooting out when you first loosen the cap, tighten it again and wait longer. If you do have water or coolant with you, when you can safely remove the cap, start the engine and slowly add the liquid to the radiator until it's within 1 inch (2.5 cm) of the top.

Flat Tire

If you have a flat tire on your truck or trailer, before you begin unloading horses or jacking, make sure that your spare has air in it. If it does not, you have two choices.

The first is to call for or flag down help so you can send in the flat or spare for repair while you wait with the truck and trailer. (This is your only choice if the flat tire is on the truck.) It will be a time-consuming operation in the best of circumstances — even if the repair shop is nearby and things go like clockwork, you won't be back on the road for at least an hour or two.

The second choice is appropriate only if the flat is on the trailer, the trailer has tandem axles, the distance to the repair shop is minimal, and the road is smooth. Remove the flat tire and, driving very slowly and carefully, limp to the repair shop on three trailer tires. Put on your emergency flashers to warn people you are going slowly. Remember, *never* unhitch a loaded trailer to drive for help. Only if you have an able assistant should you unload the horses, block the trailer securely, then unhitch and drive for help.

If the flat tire is on your trailer and you have a good spare, it is usually better to unload the horses before you jack up the trailer. Otherwise, the weight of the horses could cause the frame to bend. And if the horses start moving around, the motion could cause the trailer to fall off the jack, possibly doing injury to you or damage to your trailer. Before you unload along a highway, though, you must be very confident that you or your traveling companion can manage the horse and that the horse is of such a temperament and has enough experience that he will not panic at traffic. A loose horse on the highway often ends in tragedy, so unload only if you are comfortable with the risks.

Either way — if you jack up a trailer with horses in it or if you unload the animals first — there is a degree of risk.

Emergency Procedures

- **Don't let your cellular phone give you a false sense of security.** If you own a wireless (cellular) phone, you expect it will connect with 911 in case of an emergency. However, a wireless phone doesn't provide the fail-safe protection you might expect. Wireless networks cover only about 60 percent of the country. And if you are in a remote location, especially in the mountains, the phone may not work at all. Even in complete-coverage areas, there are dead spots. Finally, if you can't tell a dispatcher precisely where you are, officials won't be able to find you.
- **Know your location.** As you drive, keep an eye out for signs, road mile markers, and bridges, rivers, and other landmarks that will help you describe where you are.
- **Ensure that your phone will work when you need it.** Use a long-life lithium-ion battery and carry an adapter plug so you can power the phone from a car battery.
- **Know what to do if 911 doesn't work.** Dial the operator, or call a friend and ask her to place the 911 call for you.

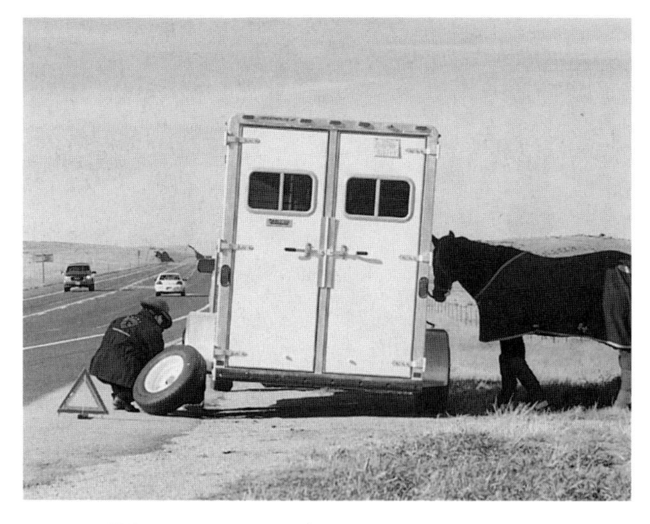

12.7 Traveling Alone

If you are traveling alone, you will have to tie the horse while you change the tire. Make sure the ground on the nontraffic side of the trailer is free of debris. Tie your horse at a comfortable length and let her relax while you work. It's nice for this horse that the trailer also is providing shade.

Preparing to Change a Tire

- If you are going to unload your horse, first pick up the highway trash behind the trailer. You don't want to add to your emergency situation by having your horse step on a broken bottle or get tangled in wire.
- Before you unload the horse, clean the manure from the back of the trailer. Even though your horse is still tied and the butt bar is fastened, any time you open a trailer door you are facing potential risk. Keep off to the side as much as possible as you work. Unload your horse according to the sequence in chapter 8.
- If you have a traveling companion, one of you holds the horse while the other jacks up the trailer or maneuvers the vehicle onto the tire-changing ramp (see photo 10.9).
- Continue to hold the horse, or tie it to the trailer, while your friend changes the tire, or vice versa.

Changing a Tire

12.8 Remove Hubcap

Before you jack up the trailer, remove the hubcap (if there is one). Use a crowbar, tire iron, or large screwdriver for this job (see page 118).

12.9 Break Lugs

Break the lug nuts free while the tire is still on the ground, but leave them snug. It is difficult to break the lug nuts free once the wheel is in the air, because the wheel turns as you exert pressure with the tire iron. The star wrench gives good leverage and you can use both hands to apply torque.

Once you've broken the nuts free, raise the trailer. You can use a trailer ramp, bumper jack, or frame jack.

12.10 Option 1
You can use a specially designed ramp to help you change a tire on a tandem axle trailer. If the rear tire is flat, drive the front tire up on the ramp. If the front tire is flat, back the rear tire onto the ramp. Here, the front tire is cradled on the ramp, and the rear tire (which is the flat tire) is in the air.

12.11 Option 2
You can also use a bumper jack to raise the trailer. The portion of the trailer where you place the jack must be of solid material: If the edge of this fender guard is made of thin metal or is rusty, it will crumple under the pressure. Note the old sock that has been placed on top of the jack to keep it from scratching the trailer if the jack leans that way.

12.12 Option 3
Another alternative is to use the frame jack from your truck. Place the jack under the trailer frame or the axle near the wheel. Each trailer manufacturer has its own recommendation on jack placement, so be sure to consult your owner's manual. If the earth is soft, put the base of the jack on a wood block. You'll need to do a bit of crawling under the trailer to get the jack positioned correctly, which is why you packed a tarp and that old shirt.

Before setting the jack in position, make sure that the jack is "down" or unscrewed all the way. Put blocks under the jack to bring it up to contact the axle or frame of the trailer — if you must extend the jack a significant amount before you even begin to raise the trailer, you might not have enough jack left to raise the trailer high enough to put up the spare.

Once the jack is in place, use the jack handle to raise the trailer.

Changing a Tire (continued)

Remove the nuts and put them in a safe place like the hubcap or a small bucket. A star lug wrench makes the job of removing the lug nuts quicker because you can spin the nuts off. (Later, it will also be handy for spinning the nuts back on.)

Remove the flat tire by pulling it off the wheel bolts and rolling it out of the way. Always be aware of traffic whizzing by.

12.13 Not High Enough
Although you have jacked the trailer up high enough to get the flat tire off, it might not be high enough to put the fully inflated tire back on. Here the spare is higher than the trailer, so the wheel bolts do not line up with the holes. Jack up the trailer until the holes and bolts are level.

12.14 Help with Heavy Tire
These big tires and wheels can be pretty heavy. A tire iron or crowbar can help you lever the tire into place. Place the tire iron in the earth under the tire and then lift up on the tire iron. This will lift and then push the bottom of the tire toward the wheel bolts.

Start the lug nuts on the bolts and finger tighten. Then use the wrench to snug the nuts up (using one of the star sequences below) just enough to seat them. Go around the star pattern again and tighten the nuts as best you can. Lower the jack or drive the trailer off the ramp and then finish tightening of the nuts with the tire on the ground. A person of normal strength can't get the nuts too tight.

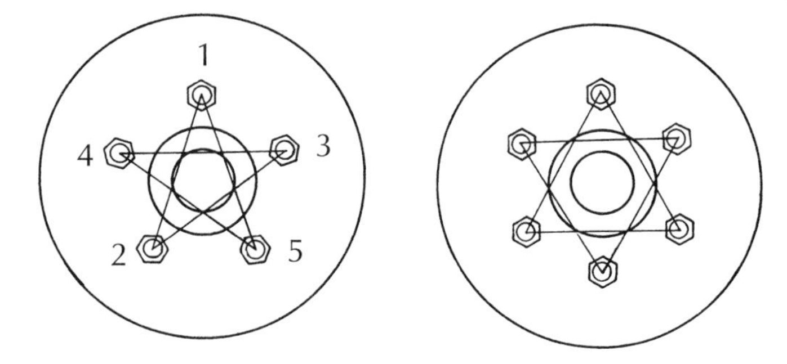

Star sequences for tightening lug nuts.

12.15 Torque Wrench

Some trailer manufacturers caution that it is essential that the lug nuts be tightened to a specific number of foot/pounds of torque for safety. This advisory is sometimes posted on the trailer and/or in the owner's manual. The only way to be sure that the lugs are at the designated foot/pounds is to use a torque wrench to finish the tightening. This specialized wrench has a gauge in the handle that determines when the required tightness has been achieved. A torque wrench suitable for this application costs between $50 and $200.

12.16 Load Up

With the spare on the trailer and the flat tire stowed away, you're ready to continue your trip. Clean up with the premoistened towelettes you packed and load your horse. Here's where thorough trailer training really pays off — nothing would be worse than being stuck in the middle of nowhere and not being able to get your horse back in the trailer. Once the horse is loaded, fasten the butt bar, close and securely fasten the rear door, fasten the horse to the trailer tie, and shut the window. Gather up your triangles. Turn off the emergency flashers. Release the emergency brake. Signal and ease into the traffic flow. You're on the road again. At the earliest opportunity, get the flat tire repaired.

12.17 Lug nuts must be tight.

Coping with Unusual Circumstances

If you find yourself miles from help on a lightly traveled road, and you've brought along a saddle and bridle and have a trustworthy horse, riding for help is a viable option. If you can, leave your traveling companion with the truck. Before heading out, just prop up the hood of the truck — the universal signal for road trouble. If you must leave a vehicle unattended, leave the hood closed, lock all doors, and tie a red cloth to the antenna.

If you're really stranded on a back road with absolutely no traffic and night is falling, you'll be glad if your tack room is big enough to stretch out in until morning. When traveling alone, choose the method of personal protection that you are most comfortable with.

Finding a Horse Motel

If you are going to travel more than 400 miles (644 km) or 8 hours, stop at least every 4 hours for at least 15 minutes. This gives the horse relief from road vibrations and a break from the muscular exertion required for balancing. A horse's heart rate lowers in a stationary vehicle. Whether you unload for hand walking or to spend the night in a horse motel will depend on the length of your trip and the nature of your horse. It might be more stressful to unload him, walk him around for a few minutes, and then reload him than it would to just let him stand on board while the vehicle is stopped.

For long trips, more than 10 hours or 500 miles (805 km), consider stopping for the night. Some RV parks and KOA campgrounds allow horses and some state parks have horse facilities. There are also a variety of horse motels (see appendix). If you are stranded in a strange town with no horse facilities, contact the local fairgrounds or county extension agent for help in finding temporary accommodations for your horse.

12.18 **Sometimes you might need to ride for help.**

Accident

In an accident involving a horse trailer, immediately call or send for police help and request a large-animal veterinarian and tow truck. Usually, it is best not to remove a horse from the trailer unless you have help. Keep the area around the trailer clear, and lights and noise to a minimum. Keep people away from the trailer. Stay calm and move slowly. Inspect the interior through the smallest opening possible to reduce the chance of further frightening a horse. If you open an escape door, a panicky horse may try to bolt through it. Consult with the veterinarian and the wrecker driver and make a plan for the best means of getting the horse out: It might be best to upright an overturned trailer without first unloading it. Often a horse will be tranquilized or anesthetized during a rescue operation. Depending on the style of trailer and its position, the horse could be removed out the back doors or through the roof once it is cut away.

Finding a Veterinarian

To find a veterinarian when you are on the road, call the Equine Connection's Get-A-DVM Service at 1-800-GET-A-DVM. You'll get the names of nearby veterinarians who are members of the American Association of Equine Practitioners. This service is available during regular business hours Monday through Saturday. At other times, leave a message and your call will be returned. You can also access the service at http://www.getadvm.com.

13 ✦ Arrival ✦

Although it might be a great relief finally to see the sign for the show grounds or the breeding farm or the public trail, you are still not quite finished with your responsibilities. Hang in there for another half hour or so, then you'll get to take a break!

You may have to park temporarily at first, so that you can go to the show or park office to obtain your stall or parking assignment and other instructions. Look for a spot that will provide shade for your horse. Open the drop-down windows so your horse gets plenty of fresh air and can put his head out. While you are checking in at the office, your horse can enjoy a look around.

Unloading

As you prepare to unload your horse, don't hurry. Remember all the effort you put into developing good trailer manners with your horse. Take your time unloading your horse to maintain the good habits he has developed. After you have unloaded and unwrapped him, hand-walk the horse for at least 15 minutes before you tie him to the trailer. This will do you both good after your trip, and will familiarize him with the grounds.

There is nothing quite so rewarding as arriving safely at your destination with a calm, cool horse. All of your preparation and care will have paid off.

13.1 **Follow the signs that indicate the contestants' or exhibitors' entrance. You don't want to pull into the wrong gate at a rodeo grounds and get stuck in a line of ticket buyers!**

13.2 **Once you get your parking assignment and are inside the grounds, there likely will be further indication as to where trailers should park.**

Trailer Parking and Tying Your Horse

One-day shows usually have a specified place for trailers to park. That becomes your horse's temporary home and your base of operations. Larger shows that offer stabling usually allow trailers to unload horses and tack near the barns and then require the rigs to be parked elsewhere for the duration of the show.

Often the parking lot is a mix of cars, trucks, and trailers. When you are choosing your parking spot at a one-day show, have respect for the person who arrived earlier than you and consideration for those who will arrive later. Allow enough space between trailers that you don't create a danger among horses, which will likely be tied to all sides of trailers. At many shows, rodeos, and public trails, there is no parking-lot attendant, so the parking is left up to the good judgment of the drivers. Try to find a space where you will have at least 12 feet (4 m) on both sides of your truck or trailer. This will allow you to safely tie your horse to either side, so you can take advantage of shade at various times of the day. To discourage someone with a small car from parking in your horse space, put a cone or plastic crate in the area.

Once the show, rodeo, or trail ride is over, remember to use the same level of preparation and care for the return trip. Don't let excitement, disappointment, or fatigue cause you to be nonchalant in your approach to trailering on your way home.

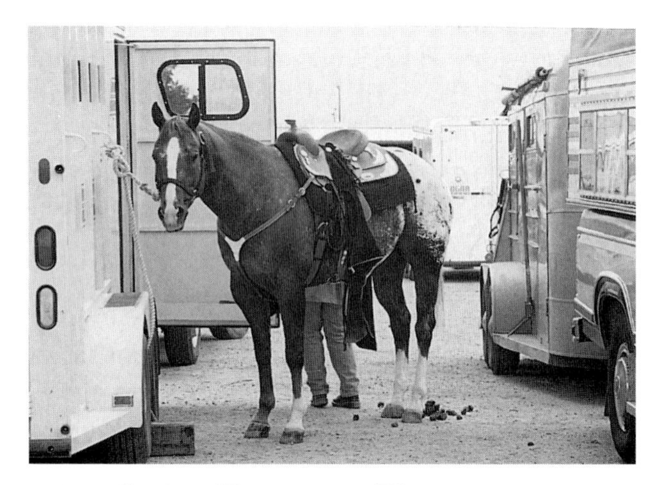

13.3 Space Between Trailers
This 8-foot (2.5-m) space would be a little too tight for many horses but this well-mannered gelding is a veteran. If a horse tried to swing around to face the other direction in a space like this, he might run into the neighbor's truck or camper.

13.4 Keep It Tidy
Keep the area around your trailer neat and tidy. This setup provides the horse with hay, grain bucket, and water bucket. When you attend an overnight trail ride, for example, plan a permanent eating station, like this one, for your horse.

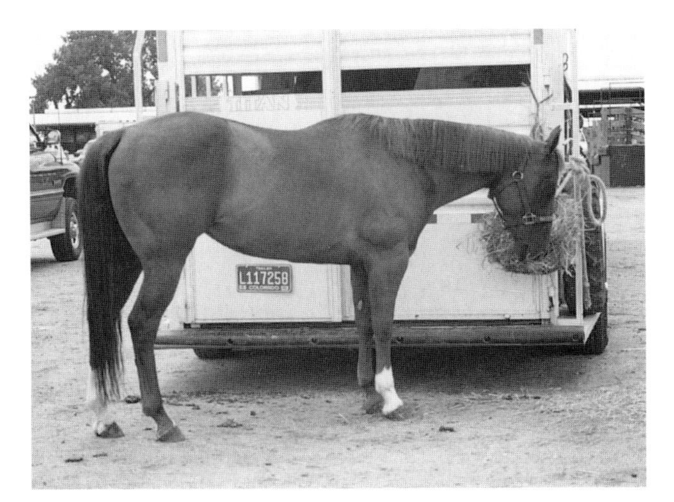

13.5 Hay Net Height

This horse's hay net is at a natural eating level right now. When he empties it, however, it could hang dangerously low, allowing him to get a pawing foot in it.

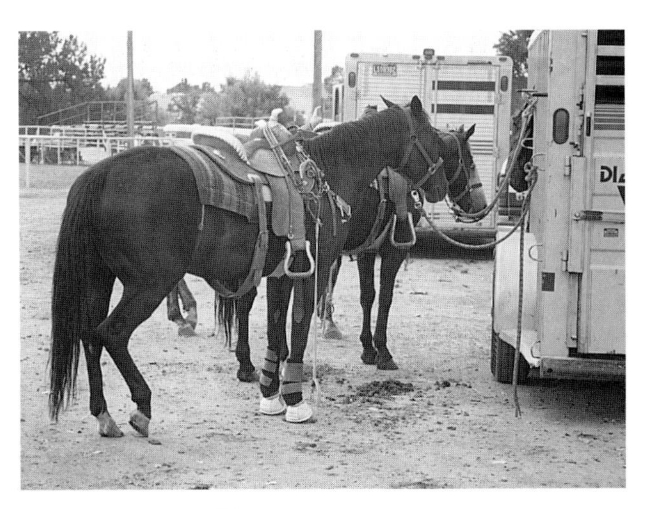

13.6 Savvy Travelers

These experienced rodeo horses tell it all. They stand three in a row, quietly, on tie ropes long enough that they can find a comfortable head position as they doze. They are obviously savvy veterans of the traveling routine. If your trailer is a ramp-load, be sure to close the ramp, as shown here, if there is a chance a tied horse could contact it.

13.7 Tied Too Long

If you tie your horse on a very long rope, it does give him room to move (this horse paces constantly) but it can also give the horse a chance to get into trouble by getting a foot over the lead rope or catching the rope on the spare tire or the bottom edge of the trailer.

Tying Tip

If you leave your horse tied during fly season, give him protection with a fly sheet and/or fly spray.

13.8 DIGGING

This young horse hasn't yet learned the patience to stand tied. He needs more work at home before he is left unattended in public. He was initially tied at a good length but now that he has dug a hole by pawing, the rope is too short when he stands down in the hole. Also, he can't reach his water pail except with his hind feet when he swings around. A pawing horse is hard on public facilities, and the repeated concussion can damage his legs and hooves.

13.9 TYING ERRORS

These two horses are a wreck waiting to happen. They are tied too low, too long, too close to each other, and by their bridles instead of by halters. Also, for safety, stirrup irons should always be run up on English saddles when not in use. Only luck and good temperament have kept them out of trouble so far.

13.10 TIED TOO CLOSE

Although these horses are tied by halters, they are tied too close to each other. Can you see what will happen next? Either the white horse will keep walking forward and cause the dun horse to pull back or one of them will get kicked or stepped on.

13.11 Adapting to Conditions

You don't have much choice when it comes to the footing of the parking lot. Here the sandy soil and the hoof-high weeds make putting on hoof black at a show a bit of a challenge, yet this young lady and her horse are obviously old hands.

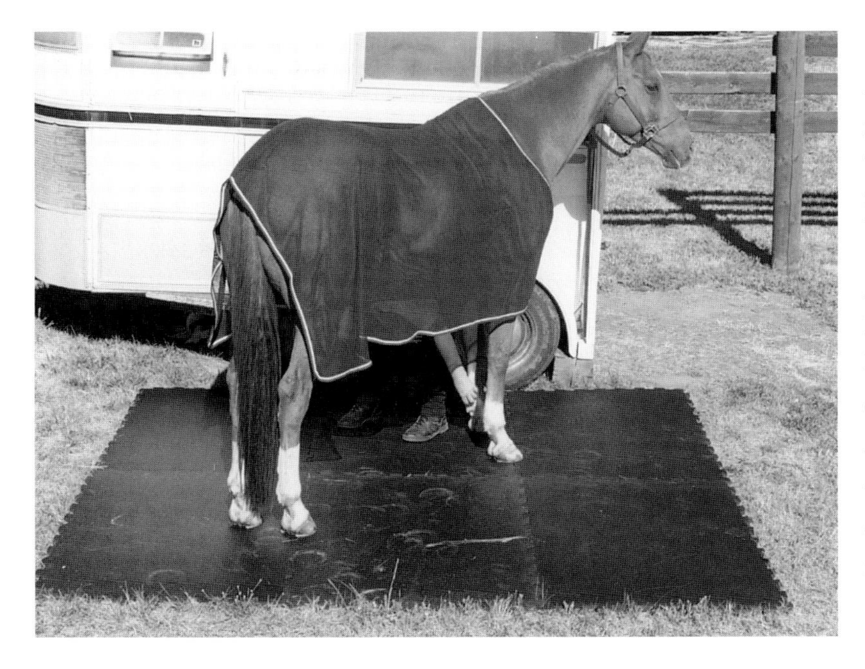

13.12 Marvelous Mats

An excellent way to make a tidy grooming area outside your trailer is to carry along specialized lightweight mats that lock together tightly like big puzzle pieces. Each 3- by 4-foot (0.9- by 1.2-m) Mighty Lite (Summit Flexible Products, Ltd., Buckner, KY) mat is ⅞ inch (2.2 cm) thick but weighs only 11 pounds (5 kg). These portable mats are also very useful for covering the stall floor when you stable your horse at a show or event. (See the appendix for more information.)

14

✦ After Every Trip ✦

If you take good care of your horse tack, truck, and trailer after every trip, not only will they last longer but they will be ready at a moment's notice to be on the road again. Make a checklist of the tasks you need to perform when you return home. If you are a frequent traveler and for convenience, leave certain items such as grooming tools and fly spray in the truck and trailer, especially if you have others to use at home.

Post-Trip Musts

- Check over your horse very carefully for anything out of the ordinary in his vital signs, skin, legs, hooves, or behavior.
- Hand walk your horse or turn him out for free exercise.
- Clean mangers of old hay and grain, and empty all hay bags and feeders. You don't want your horse to eat dusty or moldy feed on the next trip. Old, damp feed can rust the manger of a steel trailer and old hay makes great nesting material for a variety of unwanted visitors.
- Sweep out the stalls of the trailer. If you used bedding, remove it all unless you are going to use the trailer in a day or two; in that case, just remove the wet bedding. But don't let this cycle go on for more than a few weeks before you completely clean the stalls and remove the mats so the floor can dry.
- Hose off the mats and floor when needed, using a push broom for scrubbing. Whenever you hose the mats, remove them so the floor can dry out (see photo 3.2).
- Check the floor for rotting, splintering, shrinking, and warping. Repair any damage immediately.
- Wash the trailer exterior as needed.
- Clean and organize the tack room. Vacuum or sweep the tack room floor.
- Remove all tack and equipment that you will need before your next trip.
- Clean all leather tack, and wash bandages, blankets, pads, and cloths as needed.

✦ Appendix ✦

Resource Guide

Call manufacturers for information and to find the dealer nearest you.

Some Major Trailer Manufacturers and Dealers

Barrett Trailers
P.O. Box 1500
Purcell, OK 73080
888-405-4050
www.barrett-trailers.com

Brenderup Real Trailers
25 Kevin Road
Midland, TX 79706
800-745-1306
www.brenderuprealtrailers.com

CM Trailer Manufacturing
P.O. Box 680
Madill, OK 73446
888-CM-TRLRS
www.cmtrailers.com

Circle J
200 N. Kit Avenue
Caldwell, ID 83605
800-247-2535

Exiss Aluminum Trailers, Inc.
Box D, 900 Exiss Boulevard
El Reno, OK 73036
405-262-6471
www.exiss.com

Featherlite Inc.
P.O. Box 320
Highway 63 & 9
Cresco, IA 52136
800-800-1230
www.featherliteinc.com

4-Star Trailers
10,000 NW 10th Street
Oklahoma City, OK 73127
405-324-7827
4startrailers.com/default.htm

Johnson Barns & Trailers
22307 N. Black Canyon Highway
Phoenix, AZ 85027
602-465-9000

Kiefer Built, Inc.
305 E First Street
P.O. Box 88
Kanawha, IA 50477
888-254-3337

Kingston Trailers, Inc.
182 Wapping Road
Kingston, MA 02364
781-585-4337
www.kingstontrailers.com

Logan Coach
870 W 800N
Logan, UT 84321
800-742-7047
www.logancoach.com

Max-Air Trailer Sales
2120 E. Lincoln Avenue
Ft. Collins, CO 80524
800-456-2961

Merhow Industries
19757 County Road 8
Bristol, IN 46507
219-848-4445

Scott Murdock Trailer Sales
3550 SCR 5
Loveland, CO 80537
880-688-8757

Sooner Trailers Manufacturing, Inc.
1515 McCurdy Road
Duncan, OK 73533
580-255-6979
www.soonertrailers.com

Sundowner Trailers, Inc.
HC61 Box 27
Coleman, OK 74732
800-998-8779

Titan Trailer Manufacturing
2306 S. Highway 77
Waterville, KS 66548
785-363-2101
www.titantrailer.com

Trail-et, Inc.
P.O. Box 499
107 Tower Road
Waupaca, WI 54981
800-344-1326

Trails West Manufacturing of Idaho, Inc.
P.O. Box 67
65 North 800 East
Preston, ID 83263
208-852-2200
www.trailswesttrailers.com

Turnbow Trailers, Inc.
P.O. Box 300
115 W. Broadway
Oilton, OK 74052
800-362-5659

W-W Trailers
P.O. Box 807
Madill, OK 73446
580-795-5571

For other brands of trailers and updates, go to The Hay Net's Trailer Section at www.haynet.net/

Tack and Equipment

Double Diamond Halter Co., Inc.
P.O. Box 126
Gallatin, MT 59730
406-582-0706
Rope halters

E-Quest, Inc.
205 Route 541
Medford, NJ 08055
800-235-3865
Trailer Aid Jack

Farnam Companies, Inc.
P.O. Box 34820
301 W. Osborn Road
Phoenix, AZ 85067
800-234-2669
Fly masks, electrolytes

Fold-A-Feeder
Box 63
Big Piney, WY 83113
800-499-3038
Hay feeder

High Country Plastics, Inc.
5118 N. Sawyer Avenue
Boise, ID 83714
800-388-3617
Water containers

Rule Steel, Inc.
21986 Middleton Road
Caldwell, ID 83605
800-769-5674
Tandem Master Jiffy Jack

Summit Flexible Products, Ltd.
P.O. Box 520
4820 Old LaGrange Road
Buckner, KY 40010
800-782-5628
www.summitflex.com
Mighty Lite Mats

Hitches

B&W Custom Truck Beds, Inc.
P.O. Box 166
1216 Highway 224
Humboldt, KS 66748
800-248-6564

Draw-Tite
40500 Van Born Road
Canton, MI 48188
800-521-0510

Reese Products, Inc.
P.O. Box 1706
51671 State Road 19N
Elkhart, IN 46514
800-326-1090

Travel Guides

Nationwide Overnight Stabling Directory & Equestrian Vacation Guide
Janice J. Nelson
Equine Travelers of America, Inc.
P.O. Box 322
Arkansas City, KS 67005-0322
620-442-8131
www.overnightstabling.com

US Stabling Guide
James D. Balzotti
5 Barker Street
Pembroke, MA 02359
800-829-0715
www.jimbalzotti.com

Guide to Interstate Health Requirements

See page 150. Contact the office of your State Livestock Veterinarian for current rules and temporary regulations.

American Horse Council
1700 K Street NW, Suite 300
Washington, DC 20006-3805
202-296-1970

State Brand and Identification Requirements

Contact your State Brand Board.

State Trailer Brake, Chain, and Other Trailer Laws

Contact your State Motor Vehicle Division or Office of Motor Carriers.

Federal Regulations

Consult the Department of Transportation, Federal Highway Administration.

General Horse Information

Haas, Jessie. *Safe Horse, Safe Rider: A Young Rider's Guide to Responsible Horsekeeping*. Pownal, VT: Storey Publishing, 1994.

Hayes, Karen. *Emergency! The Active Horseman's Book of Emergency Care*. Boonsboro, MD: Half Halt Press, 1995.

Hill, Cherry. *The Formative Years: Raising and Training the Horse from Birth to Two Years*. Ossining, NY: Breakthrough, 1988.

———. *From the Center of the Ring: An Inside View of Horse Competitions*. Pownal, VT: Storey Books, 1988.

———. *Horse for Sale: How to Buy a Horse or Sell the One You Have*. New York: Howell Book House, 1995.

———. *Horse Handling and Grooming: A Step-by-Step Photographic Guide*. Pownal, VT: Storey Books, 1997.

———. *Horse Health Care: A Step-by-Step Photographic Guide*. Pownal, VT: Storey Books, 1997.

———. *Horsekeeping on a Small Acreage: Facilities Design and Management*. Pownal, VT: Storey Books, 1990.

———. *Making Not Breaking: The First Year Under Saddle*. Ossining, NY: Breakthrough, 1992.

———. *Stablekeeping: A Visual Guide to Safe and Healthy Horsekeeping*. Pownal, VT: Storey Books, 2000.

———. *Your Pony, Your Horse: A Kid's Guide to Care and Enjoyment*. Pownal, VT: Storey Books, 1995.

Hill, Cherry, and Richard Klimesh. *Maximum Hoof Power: How to Improve Your Horse's Performance through Proper Hoof Management*. New York: Howell Book House, 1994.

Kellon, Eleanor. *Dr. Kellon's Guide to First Aid for Horses*. Ossining, NY: Breakthrough, 1990.

Lewis, Lon. *Feeding and Care of the Horse*, 2nd ed. Twinsburg, OH: Login Brothers, 1995.

Stashak, Ted S., and Cherry Hill. *Practical Guide to Lameness*. Philadelphia: Lippincott/Williams & Wilkins, 1995.

Guide to Interstate Health Requirements

Each state establishes its own rules for animals entering its borders. These requirements are often amended. We advise that you check with the state veterinarian at your destination prior to shipment. Regulations are in effect as of January 1999.

State	EIA Test Required	CVI*	Temp. Reading
Alabama	Yes (12 mos) (B)	Yes	No
Alaska	Yes (6 mos) (B)	Yes (ii, vi)	No
Arizona	Yes (12 mos) (B, L)	Yes (iv)†	No
Arkansas	Yes (12 mos) (B, C, D)	Yes	Yes
California	Yes (6 mos) (B, C)	Yes	No
Colorado	Yes (6 mos) (H, C)†	Yes	No
Connecticut	Yes (12 mos) (J)	Yes (iv)	Yes
Delaware	Yes (12 mos) (B, D)	Yes	Yes
Florida	Yes (12 mos) (B, C, L)†	Yes (iv, vi)	Yes
Georgia	Yes (12 mos) (C)†	Yes (iv)	Yes
Hawaii	Yes (3 mos)	Yes (vi)	No
Idaho	Yes (6 mos) (B, C)	Yes	No
Illinois	Yes (12 mos) (A, B, C)†	Yes	No
Indiana	Yes (12 mos) (B, C)	Yes (iv)	No
Iowa	Yes (6 mos) (B)	Yes	No
Kansas	Yes (6 mos) (C)	Yes	No
Kentucky	Yes (12 mos) (B, C, D, G)	Yes	No**
Louisiana	Yes (12 mos)	Yes	No
Maine	Yes (6 mos) (B)	Yes (v)	No
Maryland	Yes (12 mos) (B, C)†	Yes (i, iv)	No**
Massachusetts	Yes (12 mos) (B, C, D, G)	Yes (iii, iv)†*	Yes
Michigan	Yes (6 mos)	Yes	No
Minnesota	Yes (12 mos) (B, H)	Yes	No
Mississippi	Yes (12 mos) (A, C, G)†	Yes (iv, v)	No
Missouri	Yes (12 mos) (B, C)	Yes †	No
Montana	Yes (12 mos) (C, L) (6 mos) (vii)†	Yes (ii, v, vii)	No
Nebraska	Yes (12 mos) (E)	Yes	No
Nevada	Yes (6 mos) (B, C, G, I)	Yes	No
New Hampshire	Yes (6 mos) (G)	Yes	No
New Jersey	Yes (6 mos) (L)	Yes	No
New Mexico	Yes (12 mos) (B)	Yes	No
New York	Yes (12 mos)	Yes (vi)	No
North Carolina	Yes (12 mos) (B)	Yes	No
North Dakota	Yes (12 mos) (B, C, E)	Yes	No
Ohio	Yes (12 mos) (A)†	Yes*	Yes
Oklahoma	Yes (12 mos) (C)	Yes†	No
Oregon	Yes (6 mos) (B, C, L)†	Yes (ii, vii)	No
Pennsylvania	Yes (12 mos) (B, C, G)	Yes	No
Puerto Rico	Yes (6 mos)	Yes (i, vi)	No
Rhode Island	Yes (12 mos) (B)	Yes (vi)	Yes
South Carolina	Yes (12 mos) (B, C, G, L)	Yes (iii, iv, v)	No
South Dakota	Yes (12 mos) (B)	Yes	No
Tennessee	Yes (12 mos) (B, D, L)	Yes	No
Texas	Yes (12 mos) (C)†	Yes (ii, iv)	No
Utah	Yes (12 mos)	Yes (iv)	No
Vermont	Yes (12 mos) (B)	Yes (viii)	No
Virginia	Yes (12 mos)	Yes	No
Virgin Islands	Yes (12 mos) (H)	——	——
Washington	Yes (6 mos) (B, K)	Yes (vii)	No
West Virginia	Yes (6 mos) (F)	Yes	No
Wisconsin	Yes (within calendar year) (C)	Yes	No
Wyoming	Yes (12 mos) (B, C)	Yes	No

† When EIA test is required, laboratory name and address, accession number, and test date with results must be included.
* Certificate of Veterinary Inspection (CVI) filed with the state veterinarian in state of origin is required.
** Recommended.
*** Under revision.

Footnotes for EIA Testing

(A) EIA test required for equine over 12 months of age.
(B) EIA test required for equine less than 12 months of age. For age requirement, contact the state veterinarian's office. Arizona, Florida, Idaho, Illinois (no test is required for equine under 12 months in Illinois), Indiana, Kentucky, Massachusetts, North Carolina, North Dakota, Nevada, Oregon, Pennsylvania, South Carolina, Washington, and Wyoming: No pending EIA test allowed.
(C) Suckling foals accompanying EIA-negative dams are exempt. Florida, Georgia, Idaho, Illinois, Indiana, Kentucky, Nevada, North Dakota, Oklahoma, Oregon, Pennsylvania, Texas, Wisconsin, and Wyoming: No pending EIA test allowed.
(D) EIA test required within 6 months for sale or auction.
(E) EIA test required for equine from certain states. For specific states, contact state veterinarian.
(F) Twelve months if state of origin has a state EIA program.
(G) Test chart must accompany animal. Some states require original copy.
(H) Permit required if EIA test is pending when horse is shipped.
(I) Permit and EIA test required for National Rodeo Finals. Nevada: No permit required.
(J) EIA test within 60 days if going to public auction.
(K) Oregon horses exempt.
(L) EIA test required for equine over 6 months of age.

Footnotes — Certificate of Veterinary Inspection (CVI)

(i) Pre-approved CVI from state of origin required prior to shipment.
(ii) Permit from the state of destination is required prior to entry. Texas: slaughter horses only.
(iii) U.S. origin CVI, endorsed by a USDA-approved veterinarian, valid 30 days from date of inspection.
(iv) Complete description of horse including brands or tattoos.
(v) Approved copy of CVI must be submitted to state veterinarian's office after entry.
(vi) State has requirements regarding vaccinations, testing, or other.
(vii) Six-month CVI and permit available to reciprocal Western states of California, Idaho, Montana, Nevada, Oregon, and Washington.

Index

Note: Numbers in *italics* indicate an illustration; numbers in **boldface** indicate a chart.

Other Storey Titles You Will Enjoy

Horse Care for Kids by Cherry Hill. Beginning with how to match the right animal with the right rider and progressing through feeding, grooming, stabling, health care, safety, and much more, this book provides everything a young equestrian wants and needs to know about horses. 128 pages. Paperback. ISBN 1-58017-476-0.

Horse Handling and Grooming: A Step-by-Step Photographic Guide by Cherry Hill. This user-friendly guide to essential skills includes feeding, haltering, tying, grooming, clipping, bathing, braiding, and blanketing. The wealth of practical advice offered is thorough enough for beginners, yet useful enough for experienced riders improving or expanding their skills. 160 pages. Paperback. ISBN 0-88266-956-7.

Horse Health Care: A Step-by-Step Photographic Guide by Cherry Hill. Explains bandaging, giving shots, examining teeth, deworming, and preventive care. Exercising and cooling down, hoof care, and tending wounds are depicted, along with taking a horse's temperature, and determining pulse and respiration rates. 160 pages. Paperback. ISBN 0-88266-955-9.

Horsekeeping on a Small Acreage: Facilities Design & Management by Cherry Hill. The essentials for designing safe and functional facilities, whether you have one acre or one hundred. 192 pages. Paperback. ISBN 0-88266-596-0.

Horse Sense: A Complete Guide to Horse Selection & Care by John J. Mettler Jr., DVM. Covers selecting, housing, fencing, and feeding a horse, plus immunizations, dental care, and breeding. 160 pages. Paperback. ISBN 0-88266-545-6.

Safe Horse, Safe Rider: A Young Rider's Guide to Responsible Horsekeeping by Jessie Haas. Beginning with understanding the horse and ending with competitions, this book includes chapters on horse body language, pastures, catching, and grooming. 160 pages. Paperback. ISBN 0-88266-700-9.

Stablekeeping: A Visual Guide to Safe and Healthy Horsekeeping by Cherry Hill. Explains essential features of the barn, stall, tack room, and storage, work, and turnout areas; recommends tools; and offers valuable information and helpful tips on sanitation and pest control, healthy feeding practices, safety, and emergency precautions and practices. 160 pages. Paperback. 1-58017-175-3.